CHRIS PACKHAM'S
WILD SIDE
OF TOWN

CHRIS PACKHAM'S
WILD SIDE OF TOWN

GETTING TO KNOW THE WILDLIFE IN OUR TOWNS AND CITIES

BLOOMSBURY

LONDON · NEW DELHI · NEW YORK · SYDNEY

Dedicated to Clash City Rockers with kids.

First published in 2003 by New Holland Publishers (UK) Ltd
This edition published in 2015 by Bloomsbury Publishing Plc
50 Bedford Square, London WC1B 3DP

Bloomsbury Publishing Plc
50 Bedford Square
London
WC1B 3DP

www.bloomsbury.com

BLOOMSBURY and the Diana logo are trademarks of Bloomsbury Publishing Plc

A CIP catalogue record for this book is available from the British Library

ISBN (paperback) 978-1-4729-1605-1
ISBN (ePub) 978-1-4729-1606-8

10 9 8 7 6 5 4 3 2 1

Printed in China

MIX
Paper from
responsible sources
FSC
www.fsc.org   FSC® C008047

Photographs appearing on the cover and prelim pages are as follows:
Front cover: Shutterstock
Back cover: Shutterstock
Opposite: Robin nesting in teapot: Peter Smith.
Page 1: Moorhen: A. Dennis.
Page 6: Coltsfoot growing on wasteland: Oxford Scientific Films/Mike Birkhead.
Page 7: (From the top, left to right) Common Toad: Bruce Coleman Collection (Kim Taylor); Small Tortoiseshell Butterfly: Oxford Scientific Films (G.A. Maclean); Robin: Dr Ian D.C. Shephard; Grass Snake: Chris Packham; Red Fox: Bruce Coleman Collection (Jane Burton); Hedgehog: A. Prescott; Fungi growing from pavement: Chris Packham; Garden Spider: Steven J. Brookes; Badger: Dr Ian D.C. Shephard.

THE WILDLIFE TRUSTS

The Wildlife Trusts are the UK's largest people-powered organisation caring for all nature - rivers, bogs, meadows, forests, seas and much more. We are 47 Wildlife Trusts covering the whole of the UK with a shared mission to restore nature everywhere we can and to inspire people to value and take action for nature for future generations.

Together we care for thousands of wild places that are great for both people and wildlife. These include more than 760 woodlands, 500 grasslands and even 11 gardens. On average you're never more than 17 miles away from your nearest Wildlife Trust nature reserve, and most people have one within 3 miles of their home. To find your nearest reserve, visit **wildlifetrusts.org/reserves**, or download our free Nature Finder iPhone app from the iTunes store. You can also find out about the thousands of events and activities taking place across the UK – from bug hunts and wildplay clubs to guided walks and identification courses – on the app or at **wildlifetrusts.org/whats-on**

We work to connect children with nature through our inspiring education programmes and protect wild places where children can spend long days of discovery. We want children to go home with leaves in their hair, mud on their hands and a little bit of nature in their heart. Find out more about our junior membership branch Wildlife Watch and the activities, family events and kid's clubs you can get involved with at **wildlifewatch.org.uk**

Our goal is nature's recovery – on land and at sea. To achieve this we rely on the vital support of our 800,000 members, 40,000 volunteers, donors, corporate supporters and funders. To find the Wildlife Trust that means most to you and lend your support, visit **wildlifetrusts.org/your-local-trust**

The Wildlife Trusts
The Kiln, Mather Road,
Newark, Nottinghamshire
NG24 1WT
t: 01636 677711
e: info@wildlifetrusts.org

Registered Charity No 207238

wildlifetrusts.org

Find us on
Twitter – @wildlifetrusts
Facebook – facebook.com/wildlifetrusts

CONTENTS

INTRODUCTION

Most of us live or work in cities, large towns, small towns or in areas where our impact upon the landscape is impossible to disguise. Ours is the realism of the Twenty-first Century; it is not utopian; it is not unspoiled; it is far from perfect, but none of this need stop us from finding a little romance in the wreckage.

For all our disregard for our fellow creatures, for all our negligence and selfishness, we have not managed to stifle Nature's extraordinary tenacity, her overpowering drive to survive at any cost. Thus, our office blocks and gasworks are far from sterile. They offer a resource, albeit modified, to whole communities of other animals that have found the concrete and corroded cast-iron comfortable.

It is part of a naturalist's psyche to daydream of great wildlife-laden wildernesses, of a reedbed stretching unbroken into the distant haze of a soft blue horizon and gently brushed into waves by the summer's breeze. Three Marsh Harriers flop back and forth, brushing the reed tips, peering down for a glimpse of vole or frog or fledgling. Somewhere a Bittern is booming and closer a gang of Bearded Tits 'ping' from behind the screen of a million waving stems… but stirred from our reverie, it is typically the slab side of an office block or the faded grandeur of a rusted gasworks that meets our gaze.

Let's stop dreaming and open our eyes to the wild side of our towns. You'll find things you could never have dreamed of living alongside you and, if you take a second and more studied look at the familiar, you'll see they're not so contemptible after all. And, for myself at least, the contrast between our crude constructions and the plain beauty of nature is often striking, and only serves to exaggerate the latter. A Blackbird singing in a forest is somewhat lost in the splendour, but perched on a street light above grid-locked traffic, it becomes a poignant jewel and its well-known phrases attain symphony status. So snap out of your wistful wishing for wilderness perfection, wind down your window and drink in the rich reality of a simple song. You may have overheard it a thousand times, but I bet it can still stir you, until the driver behind has to with his or her rude horn!

Left: The city may not be a romantic place, but it's where most of us live, and, if you take a closer look, you will find it is bristling with real beauty.

About this book

Not all the creatures that live alongside us are as well liked. Many, it seems, are shunned simply because they do successfully live with us and successfully resist our attempts to discourage or destroy them. In fact, the most urban of our fauna are rarely loved: rats, mice, pigeons, cockroaches and spiders are, in the main, persecuted along with weeds, whilst mosses and lichens are overlooked and ignored. Against all this prejudice I have taken the ambitious and audacious step of championing this posse in the pages that follow. I don't expect thousands of recruits for the 'Cockroach Club of Great Britain' or the 'Black Rat Re-introduction Project' nor any enthusiasm for the inauguration of the 'Royal Battersea Moss and Lichen Show'. What I hope is that in the light of new knowledge, here provided, you will consider their cases fairly, remove your black caps for sentencing and embrace a little of the 'live and let live' philosophy that I extend to these remarkable survivors.

Another important, and perhaps more pleasurable aspect to follow, is the celebration of a number of key conservation success stories which emanate from our urban endeavours. All too often, such tales are of concern, disaster and doom while there are notable triumphs blooming unseen – the return of the Peregrine and future of the Black Redstart being two perfect examples. I'm afraid that cats take their usual battering, especially in the light of new evidence, which reveals even more guilt dripping from their velvet claws. Further pleas for responsible cat keeping follow with invitations to participate in our efforts to ensure that the vulnerable and fragile refugees that remain in our cities and towns survive for future generations to discover and enjoy.

Finally, reading is one thing, television another, but there is no substitute for the real thing – getting to grips with wildlife in person. And when it is in your own backyard, there really is no excuse not to don the wellies, pick up the binoculars and turn a stroll into a city safari.

The last sections of the book provide a field guide to a number of frequently encountered species and a UK site guide. Neither are meant to be comprehensive but they do provide a tailor-made start – suitable locations for rambles and species to spot. But before you rush off, uncovering a little history of the origins and forms of any habitat often helps to further understand how each habitat provides a resource for the species which live there. So where did cities come from and how do such obviously un-natural places offer anything to animals other than humans?

*Above: Rows of argon stars line our streets each night, and as the red glow passes to orange, Blackbirds, **Starlings** and Robins make them their song posts. In the winter, daylight is prolonged by this atificial light and the birds' serenades continue into the early hours.*

Left: House Spider. Legions of animals come in search of simple shelter and, unsurprisingly, find aspects of our homes just as comfortable as we do. If you can't forgive them their intrusion, at least try to understand them!

HOME ALONE?

The best thing about our homes is that however clean or sterile they are – and not many are – they still provide a shelter for other species of animal. A diversity of wildlife from mammals, birds and insects to flora make our homes their abodes, too, but only a very few of the candidates likely to make an appearance are illustrated here. In fact, we all live in little, private zoos – albeit on a smaller scale – and as a child, I regularly led my sister on a 'hands 'n' knees safari' around our small detached house.

You see, our homes offer shelter from the elements, a fact appreciated by many other species. Fill these shelters with clutter to hide amongst, and food, and we shouldn't be surprised that a host of homeless creatures arrive at our hotels. Sadly, most people forget that we are animals and should live harmoniously with the others in our community. Thus, many of these visitors are regarded as unwelcome squatters. So try to be realistic. If they threaten to smash up the furniture or ruin the carpet, politely ask them to leave, or get down to their level and be fascinated by this tenacious group of lodgers. The wildlife on your TV may be big, brash and beautiful, but that hiding behind it is yours to explore.

① TWO-SPOT LADYBIRD
The lucky ladybird it is not. As a larva, it is a voracious predator of aphids and as the charismatic adult, it brazenly advertises its unpalatability to all. Look after these little beetles because they're the gardener's best friend.

② HOUSE MOUSE
Our loathing for these despoilers of food means that most people never see 'Mickey' alive, only mangled on the cat's mat. Their small ears and grey pelage distinguish them from the now common, suburban Woodmice.

③ RED FOX
A most beautiful beast, the fox has received much negative press – cat killer, spreader of rabies and mange. It's all nonsense. This success story is still the sexiest thing in the city despite wheelie bins denying the fox its scavenged diet.

④ PIPISTRELLE
Suburbia supports a few of these loft-loving tea-trays, found in older houses where gaps in the tiles or eaves allow access for summer roosts. This bat prefers parks and to be near water.

⑤ SILVERFISH
Three tailed, fragile and dark-loving, this is a carbohydrate feeder with a taste for flour, cardboard and old fashioned book-binding glue, but not today's synthetic stickies.

⑥ WOODPIGEON
This shy, beautiful songster has recently reappeared in our gardens in growing numbers. Whether it's due to a new found confidence or a lack of food in the countryside, enjoy them and their cosy cooing.

⑦ HOUSE SPIDER
They're attracted to your home because it's dry, warm and probably has enough wasps, flies or woodlice to keep them happy. The big females are impressive predators and capable of scuttling at four metres per second.

⑧ COMMON WASP
Art Deco masterpieces, these much maligned and misunderstood creatures are beautiful as are their papery nests. They are killers with a sweet tooth and easy to live with if you give them a break and let them get on with their job.

⑨ VIOLET GROUND BEETLE
Come the morning, this beetle can be found on its back having lost a night of fighting. This armoured jewel and its larvae are vicious carnivores, but the beetle is rarely seen in action as it is nocturnal and secretive.

⑩ IVY
Ivy doesn't strangle trees, but shelters and supports a range of insects with the nectar from its late flowers, birds, with its berries, and is the perfect home for Holly Blue Butterflies. Leave it be or plant some more of this shade tolerant specialist.

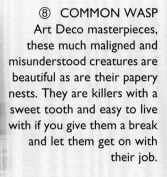

11

reasons for this development and some equally obvious consequences.

The abandonment of hunting and gathering, and the evolution of early agriculture, whereby wheat, barley, oats, lentils and peas were sown in fields, combined with the selective domestication of pigs, goats, sheep and cattle meant that people stopped moving to find food. They needed to stay in the same place to tend the fields until the harvest, and because a large labour force was required for such work, this led to the development of the first permanent villages. As a consequence of this, settlement 'societies' made up of rich and poor people, tradesmen, farmers, hunters, soldiers and governors began to develop. Because this system and its newly found stability succeeded, farming soon spread, reaching Britain five thousand years ago. Aided by the discovery of metals during the Bronze and Iron Ages, the whole environment was modified. Forests were felled, clearings created and fields furrowed, changes which facilitated a rapid rise in the human population. This increase led to all-too-familiar pressures and inevitably to feuding, then in turn to the building of fortresses. Thus, three and a half thousand years ago, the first three- or four-storey, defendable buildings had been erected. The Mesopotamian and Egyptian civilizations subsequently produced architecture the likes of which we shall never see again and the first of the world's true cities lined the banks of the Euphrates, Tigris and Nile.

Above: A Daisy pulls it off! In spite of the surrounding hostility, this little herb has rooted, leaved and flowered. What tenacity! You have to marvel at this fragile little totem with the drive to survive!

Right: It might be at the end of its beak but this Coot cannot see the ugliness of litter. To the bird, it's just available fabric to build a nest. In fact, the modern bird prefers lightweight, durable and colourful plastic above traditional, rustic materials.

The evolution of the manscape

About half a million years ago, the human prototype *Homo erectus* began building shelters. At Terra Amata in southern France, the remains of oval huts, albeit crude, have been uncovered. The stone slab floors of these were probably covered in animal skins and the roof made of branches cut with flint hand-axes. But *Homo erectus* had mastered fire so these temporary hovels would have provided an entirely new habitat – an enclosed and, significantly, a warm environment. However, this 'housing' fad was slow to take off and for thousands of years primitive man continued to choose caves over canvas. Eventually, about twenty thousand years ago, the tented inhabitants of southern Russia started living together in groups of up to sixty with several families sheltering under one roof. And, finally, thirteen thousand years later much of the human race had begun to aggregate on a permanent basis. There were some very simple

The modern city as a habitat

Tempted as you might be, I must ask you not to give way to nostalgia. Do not waste your time imagining the charms of ancient Babylon or the splendours of Thebes, Luxor or Abu Simbel. Instead, visit any city at night and with a fairer eye reassess the aesthetics of the modern metropolis.

A sparkling necklace of red, gold and green gems trickles down a gutter into a thirsty drain beneath a traffic light. The tarmac glistens, silver rain drips from gutters, neons flicker, even the litter colours the confusion and combine to produce a quirky kind of beauty. Cities are complex spaces, not flat or barren like a sheet of shingle nor as sour as a bleak winter's heath. They have a greater spatial complexity which envelopes us: tower blocks rise high over our heads and tunnels and subways swallow us. All the spaces are filled with light, sound and smell.

This 'manscape' displays a huge diversity of form. Stone, plastic, wood, metal and glass are fused into millions of shapes and textures, all welded into the massive matrix that forms the structural city. With so many units and such complexity it is little surprise that cities are said to live, breathe or sleep. Of course, this is nonsense. We are the essential components to the city's dynamic functioning. Without our maintenance and mending, cities simply decay, as Babylon and Thebes proved long ago. This is not to say that cities do not have life. They do. They support a diverse and exciting fauna and flora – the subject of this book. But how have these creatures managed to adapt to such an ostensibly alien and often hostile environment?

It is almost impossible to classify the urban landscape into a definite habitat or habitats. For instance, 'wasteland', itself a classic misnomer, typically describes an urban area of disturbed ground where Buddleia runs riot amid rubble, corrugated iron and poster paper ripped from hoardings.

Above: *Architectural 'cliffsides' and a series of 'mountain peaks' rise from the tarmac. Here, above the noise of the traffic, birds such as Peregrines and Black Redstarts find their nesting sites.*

WATERWAYS

Canals, rivers, streams, puddles or ponds offer wet conduits, which run from the countryside right into the city centre. As a challenge, I was once abandoned on the side of the M25 and told to reach the Serpentine in St James's Park without leaving water. I boated, waded, swam and paddled, and twelve hours later, after numerous natural distractions, I made it! The network that I explored is not unique. Largely due to the legacy of canal transport, many cities enjoy the same aqueous access to greener landscapes. Travellers, from fish to frogs and dragonflies to ducks can deliberately or accidentally wend their way to the wet parts of our manscape. In the recent past, pollution was a real handicap for all but the most resilient types, but now many of our urban waterways are remarkably clean, having been re-opened and restored. Some animals, particularly migratory birds or insects, actually follow watercourses because they define parts of the landscape, and are perhaps part of a map, which the animals have learned. More typical travellers include waders, waterfowl, wagtails and south-bound Swallows, which stream by each autumn. And crabs, Dabs, Cormorants and mullet can also be found in the brackish waters of the city.

① STICKLEBACK
These big-eyed, pot-bellied spiny-backed gems thrive in clean, aerated aquaria where blushing males will fight and build nests for the sober females (seen here). Their antics are difficult to observe in the wild where they should remain.

② COMMON FROG
Quick, camouflaged and not as common as it was, the frog stirs in spring to fill little pools with balls of spawn. For the rest of the year the adult is hardly seen, but the tadpoles and froglets provide food for everything from beetles to birds.

③ BROAD-BODIED CHASER
Electric blue and super fast, this jewel from the dinosaur age whirrs over water and preys on smaller insects. The larvae make 'aliens' look meek and anything goes on this murderer's menu. The best reason to picnic by water – superb!

④ MOORHEN
Boring, annoying… Even novice birdwatchers tend to disregard this neatly drawn and colourful species. This is a shame because its easily-observed private life is one of extreme behaviour, ranging from parental care to cruelty and murder.

⑤ GREY HERON
Herons are unmistakable and often seem larger than they at first appear. They are apparent, stand still in the open and do battle with eels. On misty mornings a row of five hunched in the park-pond shallows is a sight to behold.

⑥ COMMON EEL
This fish is one of the planet's top ten species. A champion of migration, it swims from the Sargasso Sea to its home stream to spawn and then returns in 'elver form'. It can go overland, eats anything and occurs almost everywhere.

⑦ BROWN RAT
This rat remains along much of our waterways. Left alone these omnivorous cleaners are an asset, but insane fear of disease and attack mean that these lovely rodents are harried relentlessly, even when they pose no threat to our lives or sanity.

⑧ MARSH MARIGOLD
This rich yellow perennial flowers from March through to July and occurs on a whole range of sodden soils. It tolerates shade and brightens pools filled with supermarket trolleys, attracting a range of insects with its pollen and nectar.

⑨ BULLRUSH
Alias Greater Reed-mace or Cats-tail, this perennial tolerates moderate levels of pollution allowing it to survive in the less than sparkling city slicks. It likes silty edges of ponds, lakes and canals. The seeds make a great mess in winter.

⑩ KINGFISHER
Blue and dashing, it sits so still in the shade of the bank that you don't spot it until it's disturbed and then all you see is a blurred missile. Frequent on urban waters it's the lack of undisturbed nesting banks which limit the Kingfisher's spread.

Right: *The crumbling ramparts of Dunbar Castle may not please the folks at Scottish National Heritage, but these* **Kittiwakes** *literally have a wail of a time each summer. This is a creative cultural co-existence from my perspective.*

Below: *Sinner in the bin.* **Grey Squirrels** *may not be everyone's favourites but let's face it, if this was a rat, most people would be running to the council. Squirrels get away with murder – rats just get murdered, and that's prejudice.*

But for all this flotsam it is the nature of the soil, its stability and accessibility to colonists, which will ultimately characterize the flora and, in turn, the fauna. Even inner city sites can be fairly rich in wildlife if they are close to corridors such as railway lines, roads, riversides or allotments, along which seeds can travel.

So really there is nothing unique about the urban habitat, only the mix of species that arrive and adapt to live there. It is an environment that offers similar niches to those found in the countryside. For example, the bare face of vertical concrete, among the most hostile substrates on offer, fairly closely equates to naked natural stone and, when used to fabricate buildings, almost perfectly replicates a cliff-side. Thus, it provides a similar resource for those species that have evolved to exploit this natural habitat and can overcome the accompanying obstacles that we impose, such as shade, disturbance and pollution. For

Left: *These apples are past their sell-by-date, but are still good enough to tempt the normally shy **Bank Vole** out from beneath its rusty shelter behind the shed. You could share a garden with a family of these mammals and never know it.*

this reason, lichens, mosses and pigeons cross the rural-urban divide and prosper. Elsewhere in the city the greener patches of our parks and gardens will conform to some mix of fragments from the rural ideal, and similarly extend an invitation to those tenacious species that can tolerate our manmade adversity.

But before we begin to explore the city space, we should recognise that we are a resource ourselves. So let us first get acquainted with a few creatures that live closer to home. Indeed, creatures that consider us as a resource. You see, each one of us is a mobile home and a provider of food for many hungry mega-millions!

The crawling skin

Soaps, disinfectants, deodorants, cleansers, wipes, antiseptics, toothpastes, bleaches and biological powders are our armaments against the evil 'germs' which, from childhood, we are told are out to get us. Our mothers and the manufacturers would have us believe that without this toxic arsenal we'd soon be a seething mass of 'germs'. But, despite the millions of pounds we pay to stay clean, that is exactly what we are!

The human womb is a sterile environment, but the process of birth is a dirty one and babies less than a day old have a blossoming population of bacteria coating their fresh receptive skins. Over six thousand bacteria will be rioting in each armpit within the hour and, in spite of our precious care, nine days later, there will be eighty-one thousand of these single-celled animals enjoying these moist zones. You see, trying to resist bacteria is a useless fight: we inhale sixty thousand with every breath due to the fact that there are as many as five million per cubic metre in the air. Our adult groins and armpits each attract a million bacteria, our scalps attract half a million and our foreheads, two hundred thousand; our backs are home to fifty thousand per square centimetre and our forearms a mere eleven thousand. And when I say our, I mean *our*, not a dirty, unkempt medieval vagabond, but a well washed, clean-living, twenty-first-century western European, for instance – *you*!

Enjoying your bacteriological buddies or simply ignoring them is relatively easy – they are invisible – but I'm afraid the next fauna under the microscope may be a little more difficult to forget about. Typically, the larger skin-sharing organisms that enjoy the human body repulse most of us and I'm afraid that, as with bacteria, we all have them.

ON THE VERGE

Roadside verges and railway embankments or cuttings are busy, dangerous and noisy, yet almost invariably undisturbed. Whilst the former tend to fizzle out in suburbia, the latter lead right into the heart of our cities. From the wrong side of the window, I've seen rabbits pursued by foxes, Roe Deer, and a great list of birds including Buzzards and Barn Owls. In truth, it's the terrestrial travellers that need this network more and for most mammals it is probably the safest route to town. Although, there are casualties. Badgers' wet bellies prove fatal on the third rail and dazzled deer and owls are occasionally hit at night. Unfortunately, the persistent 'leaves on the line' has led to over-zealous clearing of trees and bushes, and this is not good news for wildlife. However, the plants do prosper and primroses, Cowslips and Bluebells flourish before it all grows back. Plenty of species live here including great swathes of nettles, which attract modest clouds of butterflies, brambles are home to thrushes' nests and millions of mice, voles and shrews scurry about. The drains at the clinker track edge are the longest ponds in Britain and the bug-beetle-amphibian community is grateful for this living, feeding, spawning stream.

① RABBIT
'Big Ears' was once a common sight and its demise has had a profound effect on other wildlife populations. Unfriendly farmers and nasty diseases have forced this once resilient survivor to eke out a living upon sidings and verges.

② ROE DEER
Shy and secretive, you are unlikely to see these elegant creatures unless you spot their footprints. Sneaking along railway embankments keeps them out of sight and reach, and leads them into overgrown parks, gardens and allotments.

③ SLOW WORM
I hope Slow Worms still find refuge on railway cuttings, given the amount of scrap that is normally associated with them, because the present need to put flats on every last bit of 'wasteland' has robbed them of a lot of urban habitat.

④ PEACOCK BUTTERFLY
A common garden or park visitor with big 'eyes' on the wings; the adult females retreat to disturbed yards around stations which provide the stinging nettles required for legions of prickly black caterpillars.

⑤ BRIMSTONE BUTTERFLY
A splash of sulphur to start the spring. Males are regular in urban and suburban areas. Adults hibernate in clumps of ivy and live for 10 months while the caterpillars and pupae are rarely seen.

⑥ ROBIN
This vicious little thrush can be common in areas of scrub and undergrowth. Its fluid song is delivered in virtually any month of the year and is particularly noticeable in winter when most other birds are silent.

⑦ GREENFINCH
Big-billed and bullies on the bird table, male Greenfinches are fairly colourful. Their habit of singing and calling from treetops makes them conspicuous, unlike their tatty nests hidden in dense bushes of Old Man's Beard.

⑧ BUDDLEIA
It will grow almost anywhere but dry clinker or gravelly slopes that mimic its natural habitat in China are favourites. The enormous bouquet of flowers produced in summer attracts a mass of insects day in, day out.

⑨ BRAMBLE
Bramble produces delicate flowers and tasty leaves and fruits, which insects, birds, mammals and jam-makers flock to get their fill of at the end of summer. Prickly thorns ward off cats, dogs and kids and offer a nesting sanctuary for urban birds.

⑩ SILVER BIRCH
Able to grow on bare gravel or worse, it is regularly found in towns. Blown in as a tiny seed it quickly cloaks wasteland in pretty green in spring and later a wonderfully fiery autumn leaf. Birds and insects love its catkins, buds and fruits.

Head cases

Currently about one and a half million Britons have Head Lice *Pediculus humanus*, and I dare say that because infestation is seen as humiliating and shameful many more cases go undetected. As a pragmatic biologist, I see this as an unavoidable condition affecting an overcrowded society with a predilection for longer hairstyles and 'cleaner than thou' home medical snobberies. You see, it doesn't matter how clean you are, what 'class' you move within, or which shampoos you use. As we live in communities, it is highly likely that at some time an adult louse will find its way on to the fresh pastures of your own, or more often your child's, head of hair. Once its remarkably effective clasping feet have clamped it onto a shaft of hair, this wingless and flat-bodied insect will suck your blood, because this is the only meal that it is capable of eating; not your dog's blood, cat's or canary's, only *yours*.

On average these head fellows are about 3mm long, the males being slightly smaller, and each is armed with a retractable feeding tube containing three sharp stylets, which stab through the skin. Your blood pressure is enough to pump the louse's stomach full, a painless

process, which is repeated every three hours for the four weeks that the adult lives. During this time, the females cement up to two hundred and eighty eggs, or 'nits', to the hairs behind the ears or on top of the head. Fortunately, these millimetre-long, yellow-white ovoids are large enough to be combed out. However, if they hatch, the ravenous nymphs aim straight for the bloodstream. If you are blonde these youngsters become blondish in colour. In the same way, they adapt their colour to brunettes and even grey hair by the time they become highly infectious adults nine days later.

Unfortunately, lice carry one of humanity's all-time greatest ills – Typhus – and to confirm their place in the very pit of depravity, I must tell you that it is not the bite of the louse that is so dangerous but its excrement. And on that note, I think it may be prudent to skip Crabs *Phthirus pubis* because we all know how these dark red, 2–3mm long creatures are contracted. However, nowadays, this is an unlikely event because the species has become quite rare in Britain and Europe. Nevertheless, if you do encounter some unwanted pubic-pals, forget your pride and visit your doctor immediately. Do not treat yourself with petrol, fly spray, weed killer, alcohol or dog flea powder as others have done, because not only will this not work, but it will also lead to a highly embarrassing array of painful side-effects!

Invisible mite

Feeling good? Definitely no fleas, no lice, no crabs? Well, I'm confident that most of you have mites. At ⅓mm long, with eight legs and fused body segments, the Follicle Mite *Demodex folliculorum* is not in need of conservation. A minimum of one in four people, and more likely a maximum of everyone, has these sausage-shaped friends lurking in the hair follicles of their eyelashes, chin, nose or forehead. Indeed, I have plucked one of my own eyelashes and watched a *Demodex* squirm across a slide under a low power microscope. These friendly mites may live on sebum, that oily liquid secreted to condition our skin, or they may be secondary consumers feasting on the bacteria that coat our bodies. Either way, these highly successful parasites crawl head-first into a follicle and then use their strong mouthparts to hang onto the hair. Here they remain, even while mating, because males have a penis on their back so the females' underside organs are accessible without the need to release their grip on the hair. Once fertilized, the female moves deeper down into the follicle to the sebaceous gland where she lays her heart-shaped eggs. These hatch after sixty days, and after three moults and a change of follicle, they become adults that live on for two to three more weeks.

Apparently there are no harmful side-effects from the presence of *Demodex*. They do not even damage our eyelashes, and quite what threat they pose regarding infection is unknown and thus seems immaterial. Of course, they may actually be useful, perhaps controlling harmful bacteria and helping to maintain our health. But, if you have been suitably repulsed by the thought of a tiny eight-legged sausage dancing about on your eyelid, then frequent washing is the best precaution. One last thought, *Demodex* favours females because of their more frequent use of make-up. It seems they love mascara and cleansing creams! As you might expect, I'm happy to know they're there, keeping a healthy eye on my lashes!

That's the human part of our urban resource dealt with as tastefully and tactfully, albeit

*Above: One in the eye or fit for the lash? A **Follicle Mite**, seen here magnified x180, is probably probing your follicles even as you read, but don't panic. Currently, there is no data to suggest that they are dangerous and they may even be beneficial.*

briefly, as possible, so let's move on to the inanimate part of the environment that we share. What better place to begin than with our homes and dwellings; in the rooms that we make comfortable and maintain to our standards of cleanliness, where we relax or reluctantly retreat to work. Sit in your armchair, on your sofa, bench or stool and look around you. Imagine yourself getting smaller and smaller and try to envisage your surrounding space from a new and miniature perspective...

PARK LIFE

Close-cropped, sterile lawns, gaudy, non-native borders, kids' playgrounds and spilled litter bins – even the most inhospitable 'blandscape' offers the wildlife something. If nothing more, it's a home to worms, which, in turn, are food for Blackbirds, which might draw a Sparrowhawk out into the open. The fundamental building blocks for a community of animals are all present. In fact parks are often not such deserts and any aggregation of mature trees will draw in a surprisingly rich insect fauna. The nectar-full flowerbeds provide a fuelling station for the thirsty bees, flies and butterflies. Further up the food chain the trees provide shelter or nesting sites for a wide range of birds – crows, Magpies, Woodpigeons, Kestrels and Tawny Owls.

The mammal fauna is generally poor with Grey Squirrels, feral cats and domestic dogs on the ground and, maybe if there are ponds in the vicinity, a few Pipistrelle Bats in the air. The presence of water is a great bonus and, if duck-free, a park pond can claim a good diversity of aquatics such as bugs, beetles, amphibians and fish, plus all the others that come to drink. So, find a bench, don't let the pigeons see your 'sarnies' and enjoy the wildlife on offer.

① TAWNY OWL
A 'Twoo-whit' means this owl is downtown too. Where is this woodland mouser going to score a meal tonight? Well anywhere with sleeping sparrows, Starlings or pigeons probably. Listen in January for territorial disputes.

② SONG THRUSH
Let's not celebrate yet but in some places the Song Thrush appears to be bouncing back. Lunchtimes on the grass are set to be good again as this divine songster serenades us from the border or flashes us his spotty chest in the bushes.

③ BLACK-HEADED GULL
Check out the summertime chocolate cap and blood red beak and feet. Raucous? Maybe. Dangerous overhead? Possibly – but isn't it supposed to be lucky? A flock wheeling around the boating pool is an antidote to kids bickering over ice-cream.

④ STARLING
These confident birds no longer offer us the spectacular displays they once did. After persecution from councils country-wide, we're left with a few cheeky chappies pestering us for chips. If you ask me this is the 'Big Issue'.

⑤ GREY SQUIRREL
No Reds left under the bed, as the Greys have taken control. Some parks put up 'Do not feed the squirrels' notices. I asked why in Bournemouth and was told that they bite. Bite the hand that feeds you — not even squirrels would do that!

⑥ TORTOISESHELL BUTTERFLY
This common species flies from March onwards. Its larvae form tents on nettle tips but the adults flock to the nectar cafés found along the park borders. It over-winters in ivy or dry clefts in buildings.

⑦ EARTHWORM
How can this aerator of the sod be an enemy of the lawn? Its tightly-curled soil heaps apparently! Still it's a must for Blackbirds, Magpies and the rest. At night, parks and golf courses become the Badger's foraging grounds — a real urban bonus.

⑧ DAISY
This pretty flower is trod on, trampled and mowed but still its dainty, pink, fringed face grows back. From sea level to 1000m, from sodden to dry soil, from acid to alkaline, from January to October this perennial prospers.

⑨ DANDELION
This flower sets seed without fertilization. With over 200 British species, it is widespread, especially in the city where the most inhospitable crack will capture a parachuted seed and produce a glowing flower.

⑩ LONDON PLANE
A hybrid found in cities all over the UK. It has a distinctive flaking bark, which peels to produce attractive patterns. Growing to 35m in height, it did well before Cable TV brutalized its roots — an outrage in spite of Sky Sports!

THERE'S NO PLACE LIKE HOME: INSECTS

Our homes shelter a fantastic variety of species, but unfortunately, we react viciously towards all but a few. Those labelled 'pests' we detest, and yet they are amongst the most successful creatures ever to appear on earth. It strikes me as sad that most of us know more about how to kill them than the fascinating intricacies of their life histories. To change your outlook, read on…

ANOTHER WORLD

There are places in your living room where you have never been, and you are spared by your human size the horrors they harbour. Imagine you have fallen into the crease of the chair where biscuit crumbs are boulders, or you're hanging on the sheer face of the curtain at the equivalent of ten thousand metres where a mild draught is a raging gale, where a distant 100-watt light bulb is the sun, where the edge of the world you know lies beyond the plastic rail. Imagine you are in the tangled forest of the carpet pile, wading through a greasy swamp of dust, cat hairs, human skin, rotting food particles and dirt. Think of all those places you wouldn't dare to tread if you were only ½mm tall – behind the oven, under the washing machine, at the back of the bath, or in that slime-filled gully that runs along under the leaking edge of the sink unit that you have been meaning to fix for months. A horrible fate awaits you in your own apparently hygienic home, and that's without even acknowledging the physical forces which would shape your life on the micro-scale.

If you fell from the summit of the curtain, you would probably bounce happily to rest on the canopy of the carpet forest. But if you stuck a careless limb into a huge hemispherical water droplet on the kitchen work top, the force of surface tension would drag you in and drown you as it all too often does to careless flies and ants. The seemingly comfortable strands of carpet fibre would be hard, sharp, and charged with lethal pulses of static electricity. With such a distorted body surface area to volume ratio your temperature tolerance would be hypercritical. A centimetre too near the fire, and you would be as dry as dust in seconds, and another too close to the door and you would be a frozen flake in even less. All the apparent densities of objects would be radically changed, indeed, even the density of the air itself may mean that for miniaturized humans breathing would be impossible, and this alone would rule out any such fantastic voyage. To exploit your home comforts on

Right: Two-spot Ladybirds. These popular little beetles are terminators of aphid larvae. They hibernate as adults so such gatherings are common in garages and sheds each autumn.

Above: *The **Hornets'** nest is an architectural marvel – centrally heated and air-conditioned by its inhabitants, and an interior design modified over millions of years. This grossly maligned insect needs care, conservation and a little more consideration.*

Left: Peacock Butterfly. *Our 'vanessid' butterflies over-winter as adults and need a dry, cool place where they can avoid mice, spiders and mould. Garden sheds and ivy are familiar sites and their closed wings are perfectly camouflaged.*

this scale you would need a whole new body design, capable of withstanding desiccation, wrapped in a tough exterior armour with modified vision, smell, hearing and breathing. No nose, no hands and no heart, but some wings might help. With the body of an insect you could join the ranks of the domestic survivors with which we co-habit, often so unharmoniously.

Our homes, offices, warehouses and factories are designed to provide shelter, to be constantly dry, warm and still, and ultimately to be comfortable. But our requirements are far from unique and such a cosy stability is enjoyed by an

extraordinary range of refugees from the world outside our windows. The precise supporting fauna of any building depends upon the building's use, location, the building fabrics and levels of disturbance, but because these factors are somewhat limited in range, all of the usual suspects crop up in most gangs of squatters.

Wasps in the roof, flies around the lampshade, spiders in the corner, moths in the bathroom, butterflies in the garage and woodlice everywhere. Even in the heart of the city, woodlice appear in apartments, flats and offices to die on the dry carpet and rattle up the vacuum tube. To most of us these lowly lost souls have a hapless charm, born in childhood when these little tanks tickled our fingers. Thus, an inbound woodlouse is more likely to engender sympathy than a shriek. They have no jaws as such, no sting nor even long hairy legs; they simply amble about like thickly coated pensioners and appear just as harmless. Joining us indoors is a precarious judgement on their part, because straying from their familiar damp enclaves does these funky little detritivores no good at all.

CRUSTACEANS ON THE CARPET

Britain has about forty-two species of woodlouse and all are more closely related to crabs, lobsters and prawns than to insects. They are one of the few groups of land-living Crustacea that walked across the seabed, through the intertidal zone and up on to land about 160 million years ago. They were well prepared, having strong exo-skeletons to support their bodies, 14 sturdy legs for rapid locomotion, chewing rather than filtering mouthparts and internal fertilization. The development of a brood pouch to hold the developing young protects them from desiccation, so unlike all the land-living crabs, they don't need to return to water to breed. However, because they lack the waterproof skin typical of insects and breathe using modified gills that lose a lot of water, they have been damned to live in the cool of the dark and the damp. Central heating is an anathema to these creatures whose normal response is negatively

*Right: A **Violet Ground Beetle** poised and ready to rumble. This splendid insect is a ferocious carnivore and will take on anything smaller and softer than itself. A jam jar favourite, it is clad in a coat of iridescent finery.*

*Right: The big, black blackguard! Slugs such as this **Large Black Slug** are prospering in the damp warm summers and few people are pleased. I've heard that a chap in Bristol can identify slugs to species level by tasting them! If you try this at home, stand by with Listerine!*

her head with his mouthparts and then drums on her carapace with his front legs. After a full five minutes of this tender courtship, he bends his body diagonally under the female to present his left genitalia to her right genitalia. After several minutes or so, he swaps sides and repeats this manoeuvre before sloping off into the darkness. The female immediately stops feeding, moults her skin, tail end first, and eats this transparent husk to preserve vital nutrients. Simultaneously, she develops a brood pouch, which forms a false floor to her body and which fills with liquid to receive the fertilized eggs. Here they develop, until the miniature but mature-looking young crawl out and mass in densities of up to two thousand per square metre. At first, these hatchlings only have 12 legs and will need to moult twice before gaining their final pair of legs. Such densities are

phototropic – they turn away from the light and seek the dark – and positively hydrotropic – they actively gravitate toward moisture. Quite why they make such numerous suicidal forays onto our carpets is thus a mystery.

Despite this silliness, certain aspects of their lives are relatively stimulating. Woodlice mate in total darkness. When a male finds a receptive female, he tests the air with his flickering antennae before gently beginning to tickle her. If she doesn't flee, he climbs on her back, licks

Left: ***Common Pill Bug***.
*Little tanks that trundle
and bumble and without
them we'd all be knee
deep in leaves and logs.
Quite why they stray
indoors is often a
mystery, but there can
be few houses in any
city that haven't hosted
these hordes.*

designed to satisfy the safety in numbers rule. Baby woodlice are defenceless, soft and tasty – thousands perish for a few to survive. Of course, all this happens outside your house beneath loose stones and rotting logs, and there is no chance of such a plague appearing on your parquet. Whilst few people can be more than irritated by having to hoover

Left: *This distressed*
Common Pill Bug,
*perhaps just peeping out
through a chink in its
carapace, is otherwise
safe from the jaws of
predators such as the
beetle opposite. Not
all woodlice species can
do this, so don't try to
force them!*

up occasional woodlice carcasses, some other housebound hordes generate a little more concern, and as is so often in life, size is not always that important.

Right: *Why is it that the* ***Clothes Moth larvae*** *refuse
to eat that giant jumper that your mother knitted
and always head straight for your favourite woolly
warmer? Whatever, even I can't welcome these
little blighters into my wardrobe – no toxins, but the
doors are closed.*

Below: Black Ants' nest.
As soon as you expose an ants' nest with your finger, you'll see them rushing off with their precious pupae.

Above: 'Don't tread on an ant, it'll come looking for you.' proclaimed Adam Ant in the 1980s. Well, it's unlikely actually, as a **Black Ant**, such as this, is more interested in appropriating the sugary excretions of aphids than fulfilling pop fantasies. Nice idea though!

THE ANTI-ANT LEAGUE

Central heating has encouraged a multitude of creatures to come in out of the cold where they could not otherwise survive. A classic example is the Pharaoh's Ant. Obviously from Egypt, this tiny socialist arrived in 1828 and at first found shelter in greenhouses. Today, however, it is widespread in our cities where it is a nuisance in hospitals or anywhere where it can compromise what we deem to be acceptable standards of hygiene. At 2mm in length it is smaller than the familiar Black Garden Ant and different in that it doesn't make a fixed nest. It roams in search of food and moisture: in fact, it is said to roam into bandages to feed upon suppurating wounds, into baby's incubators and is allegedly attracted to patients with high fevers. I once asked a number of staff in a hospital about these charges and all confessed that they had never seen such things. Ants were the least of their problems. I then went to the loo and there, in a living line that laced across the tiles, was an Arab army on the move. These marauders have also exercised a right to roam into tower blocks where, once established in the network of heating pipes, they are very difficult to remove. Whole blocks have been evacuated for teams of 'pest' controllers to go about their toxic tasks. But the Pharoah's Ant will continue to roam on because bands of their workers will carry eggs, including those of new queens, to fresh sites to establish satellite colonies. Apparently they can spread a horrid cocktail of contagious bacteria but hospitals seem to require little help to do this anyway. (The current plague of MRSA, a drug-resistant and dangerous bacterium that is omnipresent in British Hospitals, is a case in point.) I feel blaming ants is a little cowardly. And as I've been known to ask before: how many people have died of ant?

SIMPLY THE BEST

There are species on Earth which could claim to be the greatest of all survivors. Sharks are a famous example, as are frogs, turtles, crocodiles and dragonflies, all of which survived a series of planetary mass extinctions and still exist today. But there is one creature that does better than thrive, it persists in spite of us and in spite of almost everything. It is the ultimate animal, the best, the most successful, the indestructible leader of life's gang, the most tip-top... cockroach.

Over the years I have regularly woken with a raging hangover and the inevitable ferocious thirst, and, never wishing to blind myself at the flick of the switch, have stumbled in darkness towards the kitchen or bathroom sink, desperate for liquid. Here I have received a disconcerting surprise because as I have opened doors, illuminated rooms and squinted in, I have stood bleary-eyed before the most belligerent burglars imaginable!

Right: Living off the fruits of our labour.
Cockroaches *can eat anything and live anywhere, and we don't like it at all. Apart from me – I think they're greaaat!*

and scurry they do as soon as the light is lit. Before one might arrest them they zip under the nearest edge into a gap only a few millimetres wide. And thus in hotels, hostels and hovels across the world I have stood and been repeatedly reminded that I am the subordinate, the late-comer, the flash-in-the-pan pretender, the mammal, the dominant executor of influence on Earth, who nevertheless must prostrate himself at evolution's altar, and acknowledge a better life form: the indefatigable cockroach.

Peering into the tiny dark slits where they've disappeared, I have wondered if they were only lonely males and not mated females, pregnant with an impending infestation of their clan. Indeed, as I've gulped down my refreshment I've only been able to think of one thing: being

Their heads bend under their oval and flattened bodies, their hidden mouths are capable of biting, chewing, or worse, licking. Their eyes are huge and their antenna whip-like. Their legs are slender, spindly and designed for rapid scurrying,

Left: *Another ancient animal that seems to annoy people – the dainty **Silverfish** (seen here magnified x3). This larder-loving scavenger can exist on the most meagre diet of dusty crumbs, and rarely strays out of the dark.*

beaten, being a second-rate organism, being outdone, outwitted and outlived, frankly being in awe of this extraordinary animal.

The curse of the cockroach

These insects are described as the most obnoxious creatures known to Man. Further, at home or at work they are, and I quote, 'psychologically disturbing' and 'capable of causing mental distress'. Well, I have been disturbed and distressed but not through exposure to insects and you may think I'm unbalanced even to attempt to extol the 'virtues' of one of the most loathed insects in the world, but here's my shot.

There are some 3,500 different species of cockroach, of which only one per cent are considered 'pests'. They are one of the most ancient winged insects, having evolved over 250 million years ago in the Carboniferous period. Thousands of their fossils have been recovered from coal measures, and this disproportionate abundance of a single group has even led to an alternative name for that geological period – 'The Age of Cockroaches'. The warm, moist conditions at that time led to their evolution where North America is now situated, but recently these almost unchanged and relatively unspecialized insects have spread with the help of man to nearly every corner of the globe. Currently only fleas and lice rival them in terms of the intimacy of their relationships with humans. Their speed, their apparently irrational direction of movement, their staggering reproductive potential and their habit of fouling food means that they are associated with poor standards of hygiene and universally despised. They are public

health enemy number one, and probably our most unwanted allies.

There are only three native species of British cockroach – the Ectobuids – which inhabit the dry heaths of southern England. They are small, greyish-brown insects that seldom grow larger than a centimetre in length and never attain 'pest' status. It is the Common, or Oriental, Cockroach *Blatta orientalis* and the German Cockroach *B. germanica* that present such a formidable challenge to our endeavours to improve our living standards. They regularly crop up on food counters and in hospital laundry baskets, incubators containing premature babies, beauty salons, bundles of clean linen returned from laundries and restaurants – even the best restaurants! They are able to exist in almost any form of human habitation, their firm favourites being pre-war hotels, restaurants, cafés, canteens and hospitals. They like old buildings full of nooks, crannies, crevices, hollow walls and warm pipe-ways – all perfect places, especially for breeding!

Controlling them is very difficult and in some cases, impossible. They have proved useful as laboratory animals, especially for space research and studies in insecticidal control, but at the same time they have accidentally been introduced into schools, colleges and universities across the world where they remain despite ever more diabolical chemical assaults. Because of their extraordinary ability to adapt and react to any new poison – they can even taste it before they eat it – they remain all but indestructible. Their rapid, greasy feet allow them to scurry over pipes as hot as 300 degrees Fahrenheit, they can go without food for up to three months and have a reaction time of only one $^{54}/_{1000}$ of a

LIVE AND LET LIVE

Why let aged and misplaced prejudices impinge upon your reason? You don't have to like the creatures that come in from the cold but you don't have to kill them either. It isn't difficult to outwit an invertebrate, or modify its behaviour so that it leaves, perhaps not happy, but at least with the option of bothering your neighbours. All that's needed is a little research and a little effort. Go on, make the punks' day for a change, and remember, they'll still be here when we're gone.

*Right: A bit brash and vulgar, the **Peacock Butterfly** is one of our most vigorous species and very at home in the urban environment. Nectar and nettles are all it needs to cheer up the bus queue or the playground party.*

the wall with its feet and antennae it receives a wicked 2,000-volt charge. This will kill individuals, but it won't exterminate armies, and it's the armies which continue to march.

In 1987, the flying Asian Cockroach invaded North America via Disneyland in Florida. There was nothing 'Mickey Mouse' about the effectiveness of their operation: one pair giving rise to four hundred thousand offspring in only one year. They like light, flee from insecticide and have now conquered most of the continent. From the magic castle to the mid-west, from cartoon-land to Canada, from Cinderella to California, Colorado, Kansas and Kentucky,

second, which obviously makes swatting or stamping difficult. A high voltage zapper-trap has been devised. It is plugged into the mains and if a cockroach sniffing out a small bait tablet touches

'Hello, nice to meet you, I'll be your pest today, you can call me "cocky" because I have a right to be!'

We British owe the gift of the cockroach to Sir Francis Drake, who in 1587 'took a ship' (he

*Right: The deckled edge and the varied brown of the wings make the **Comma Butterfly** a triumph of camouflage when it's at rest as here on the bramble leaf. This bush's flowers and then its ripened fruit supplies all the sugar this species needs to get up and go.*

Left: *The spotted undersides of the **Common Blue Butterfly** may lack the snazzy splash of blue found on the male's upper wings, but in close up the design is beautiful. What remarkable little machines they are! What do these antennae sense? And what is it like to fly on those fragile wings? I still wish I could be a butterfly for a minute, just to find out.*

was a pirate as well as a hero) named the *San Phillipe*. This Spanish East-Indiaman was laden with spices but also crawling with cockroaches, and keen trading soon ensured that these stowaways were landed in all the ports of old England. Once established in London, these cockroaches had become common by 1644, but then only slowly spread around the country, such that Gilbert White noted with intrigue the first cockroach to reach Selbourne in Hampshire, in 1790. Upon the high seas though, cockroaches are intrepid travellers, and on the mutinous voyage of the 'Bounty', Captain Bligh attempted some very crude methods of cockroach control. He drenched the whole of his ship in boiling water, a tiresome task and what's more it didn't work. Today, the US Navy uses tens of thousands of litres of insecticide every year to try to control cockroaches, but recently twenty thousand were found behind panels lining a ship stateroom, so this method obviously doesn't work either. Legend states that the leather of sailors' shoes and even their skin and fingernails may provide food for cockroaches enduring hard times, so not surprisingly Japanese navymen could once earn shore leave by killing three hundred cockroaches.

Such horror stories abound in the field of 'cockroachery' and one of particular note concerns cockroaches gnawing the eyebrows of sleeping children. Doubting it as obvious bunkum, a chap named Rau tested this in 1940 by sleeping in a room deliberately infested with cockroaches. He was eventually awakened by a tickling sensation on his face and spied a long pair of antenna playing over his cheek whilst their owner's extended mouthparts were 'imbibing moist nutrients' from his nostrils. What a genius! I salute him as a true biologist albeit in the old-fashioned sense, a simple experiment and a resounding and most memorable result!

TERROR BEHIND THE TELLY

The typical human reaction to cockroaches may occasionally be justified but throughout kitchens, living rooms, halls and bathrooms across the land, women, children and grown men exhibit an irrationality bordering on idiocy when they tower over an innocuous House Spider *Tegenaria* sp. Cries of 'Aaaarggghhh!', 'Eeeeeekkk!' and 'Ooooh nooooo!' are expelled with a greater frequency than they appear in *Beano* comics and unfortunately, many harmless arachnids become legless and brainless under the slippers of the thoughtless.

Despite never having feared spiders in any shape or state of hairiness, it would be unfair

Right: A nice clean sink, no scum or scale here, just the perfect glittering backdrop to show off the daredevil House Spider doing its surfing stuff. Soap and sanitation seem to be no barrier to the spiders who want to come in from the cold. Go on, offer some 'Shelter from the storm'.

Right: The beautiful Garden Spider – voluptuous, brutal and coated in a brocade straight from Henry VIII's waistcoats. The later the first frost, the fatter they get gorging on the available food. If it's urban jewellery you're after, then this is the brooch you want to find at the bus stop.

of me if I mocked the genuine sufferings of arachnophobes, after all this is a very real and common psychological condition. If confronted by the dangerous Brazilian Brown Running or Funnel Web Spider, alarm may be a life-saving reaction but British spider bashers need not live in fear for their own lives. Even Mr James Bond, an apparently well-educated super-spy, has been seen swatting virtually harmless tarantulas from his person. So why is such a phobia so common and so often exaggerated?

There is some evidence that the fear of spiders is a conditioned response, that it is innate or present at birth. Young apes and humans have been seen to produce anxious reactions upon exposure to spiders without previously learning

to fear them through parental instruction. Nevertheless, arachnophobia is undoubtedly exaggerated by the reactions expressed by one's peers and in response to these a fear is instilled and retained. The phobia is not sexually biased towards females, as is frequently imagined, as a greater proportion of adult human males show considerable anxiety – shaking, sweating and cursing. It is they who actually have more difficulty facing the timid and shy House Spider that skulks behind their precious sports-providing satellite box. Oh dear, cheering for England but they can't pick up a spider!

For a phobia of this ilk to persist psychologically the source must be frequently encountered as is the case when spiders are common and a number are fond of human habitation. Such repeated exposure leads to stress, avoidance and a lifetime of irrational fear. But this doesn't have to be the case because arachnophobia is treatable in all but the most severe cases. A course of reinforced practice procedures should enable a parent to cure a child of such a phobia. Firstly, insist that exposure to our eight-legged companions will be therapeutic. Then, show a spider to the child and praise them for their tolerance. Over a period of time make the praise more difficult to obtain so the child has not done well enough until they have touched or even handled the animals. Such systematic de-sensitization should be taken slowly, perhaps in ten to twenty steps, and before you know it your child and their friends will be bringing all their old spiders over for your collection!

Of course, there have been other concepts for cure. The novelist Sax Rohmer once wrote to a national newspaper claiming that his fear of spiders had been suppressed by eating one. Bristowe, an arachnologist of repute, was obviously partial to them. He ate one to win a game of cards and then offered to prove that a House Spider a day does not keep malaria away. Fortunately, The London School of Tropical Medicine

Left: House Spider. What an athlete! Body held high on a feat of tech-leg-gineering. Steroid-free, still a champion on the carpet and ready to break the record as the world's fastest running spider. Four metres per second and coming your way – get the medal ready!

*Left: Jaws on your carpet. The **House Spider**'s reduced limbs that arch away from the 'face' are the pedipalps, which are used for manipulating food, and mating. The fangs or chelicerae are partially hidden here but always ready to do the dirty deed on demand and always with diabolical effect.*

Right: *T-rex has decapitated Barbie and the 101st Dalmatian was too dumb to prevent the tragedy. Cleaning up after the kids will find you sticking your hands into places that you normally only push the vacuum. Places where the toys have real live friends, not the plastic Praying Mantis here recovered but a host of clandestine cohabitants.*

refused his offer of research and both he and the spiders survived.

In spidery circles, House Spiders are renowned for their speed. Adult females can sprint over 330 times their own length in just ten seconds, so a scaled up Olympic equivalent would see the 100 metre record shattered in just over a second. In fact such pacey behaviour has led to spider racing. In the 1790s, things must have been dour in Cornwall because, here, marked and trained House Spiders were released onto a hot plate in the centre of a table. The first over the side was the winner. The sport was limited to sprints because speedy though they are, House Spiders fall exhausted after only twenty seconds of eight-legged athletic pursuit.

So there, if you make it out of the room, there is no need to run up the stairs, the tired Tegenarian won't make it. Oh and one last note. Arachnophobia is not inherited. My girlfriend is severely afflicted, I am not. Our seven-year-old daughter tickles spiders for kicks and, pinned on her bedroom wall is a Polaroid of herself holding a large hairy tarantula. I'd have it framed — her mother won't even look at it!

WINGED INVADERS

More than any other type of animals it is the insects and their invertebrate buddies that will appear in our houses, often repeatedly and in numbers. Our predilection for lawns, whether in park or garden, means we create a near-perfect environment for the *Tipulids*, crane flies, so often incorrectly referred to as 'Daddy-long-legs'. Their larvae, nick-named 'leather jackets' on account of their tough skins, feed on the roots of grass and when the adults emerge into the cool of the late summer evenings they are attracted to lights and come indoors to clatter around the lampshade. We have played host to some huge ones with patterned wings — real beauties — but the trick is catching them without de-legging them, a bit tricky as they carouse and careen in drunken dives behind the sofa.

Bees and wasps are generally not too keen to come indoors, the latter preoccupied with preying on other insects outside. At the end of summer, queens might consider over-wintering behind curtains or in the shadowy security of the shelf at the back of the sideboard, but you are more likely to find them buzzing about in the

spring. See them out kindly please and if they choose to nest in your loft, don't panic, be a more understanding host and learn to live with these exquisite creatures. Wasps don't fly at night but Hornets do and they are attracted to lights. Again be tolerant, they are burdened with a grossly exaggerated reputation for violence and are not common. Simply close the windows on the side of the house where they appear to enter and put any repeat offenders out using the old 'glass and paper' technique.

Flies are another group that are much maligned and upon whose extermination we overspend. If they cannot land and feed on any food that you intend to eat, they have no real means of infecting you with any of the diseases they are said to spread. Now, I'm no fan of those lacy-lampshade-things that granny hides her cream cakes under but I'd rather keep my food in the cupboard and enjoy the wacky antics of a few House Flies around the place than race for the aerosol and nearly perish in a sneezing fit myself.

The appearance and abundance of any butterflies or moths will be directly related to your proximity to nectar as food and those plants which provide forage for their caterpillars. Thus even in the centre of our cities the Vanessids (Red Admirals, Peacocks, Tortoiseshells) will put in a fly past and some of the tree-feeding larvae of moths will pupate into those wonderful monsters that always seem to appear on pub toilet walls. Lime and Poplar Hawk Moths, the various 'under-wings' and the delicate and unmistakable Swallowtail Moth whose caterpillars actually eat ivy, all put in sanitary appearances.

Beetles can be bad news if they're wood-borers. I fondly remember evacuating the family home each summer to stand in the street watching a pungent purple smoke billow out from beneath the eaves and then returning indoors to find the whole house, and more importantly my Airfix models, completely covered in dust. It was woodworm control and apparently these tiny pellet-like creatures were on a mission to eat our house. It still stands despite the practice of lighting smoke bombs having passed without word years ago. Was it worth it, was it necessary, did it really work or was it just more psycho-pest paranoia? Weren't Kennedy and Khrushchev doing more dangerous things in Cuba? Weren't the Beatles more important than the beetles?

*Above: What a beauty! A **Lime Hawk Moth** at rest, though it must be said that, typically, they find a slightly more shaded spot to hang out for the day. When freshly emerged, their wings are like velvet, a finish that fades during the course of their month-long adult life.*

*Below: The **Emperor Dragonfly** has come from the Jurassic jungles to the concrete contemporaries. Even modern bodies of water prove tempting but water is not essential as most species are great commuters.*

Long ago a family moved to England from Europe bringing with them a cherished wooden wardrobe, which they kept in their home. Fifty years later a dead beetle was found on the

Left: A little understanding can alleviate a lot of unnecessary fear. Movement is essential to provoke a wasp to sting. So the insane flapping dances that people perform each summer are the worst reaction to these insects. That said, away from the nest, stinging is the last thing a foraging **Common Wasp** *wants to do. Such a potentially suicidal gesture would weaken the fitness of the colony by one, and that is not good evolutionary sense.*

windowsill of the room where the wardrobe stood. It was unusual and was passed to a Coleopterist who identified it as a species found only in Scandinavia where the furniture had originated. Its larvae had been slowly maturing in the wood, and unnoticed had been gnawing tunnels and chambers for at least half a century, probably longer. To me this anecdote prompts several considerations: principally that 'our time' is not the time that scales the lives of so many of the organisms with which we interact. We live in such a rush that the thought of living in a Scandinavian wardrobe for fifty years strikes us as more science fiction than fact. But then, I've seen some of the latest-fashion furniture items and

I can't imagine them lasting that long let alone providing a home for larvae. These days, for some things, living a long time is becoming extremely difficult indeed.

Left: **Common Wasp***. It's a supermodel – sexy, lean, perfect and beautiful – and a feared and fierce insect predator – the insect version of the Peregrine. Show it some respect, and imagine being nibbled by those mandibles… ooooohhh!*

URBAN CHAMPIONS: CATS AMONG THE RATS AMONG THE PIGEONS

Please, please, please don't even think the word 'pest' or 'vermin' or anything of a similar nature as you read on. Here are some animals which our activities have directly and obviously encouraged through our and their domestication. Now that they all have the capacity to become a nuisance in some way, we must recognise that we've made these monsters, and it's up to us to put things right – preferably in a positive way!

LOVED, LUCKY AND LOATHED

Cats were domesticated three thousand years ago in Egypt from African Wildcat stock and ever since then they have enjoyed mixed favour from humans. Initially worshipped and cherished as exotic pets, cats' good fortune continued in post Roman Britain as they provided fur for clerical collars and were awarded their 'lucky' status. But this luck was short lived as, during medieval times, they became associated with black magic and the supernatural and were widely hunted and tortured. The currently cosy age of the Domestic Cat as companion and pet was heralded in the late 1800s when the first cat shows were inaugurated and interests in breeds and all things pedantic and pedigree began. Today Domestic Cats are an inescapable fact of life in our cities and suburban areas, pampered, protected and probably responsible for more direct damage to our wildlife than every other negative factor. Needless to say, I don't like them!

The cat flap!

My rationale for despising cats is reasonable because it is purely ecological. They are super-predators and the impact they have upon our urban wildlife is clear to see. The Mammal Society's 'Look what the cat's brought in' survey conducted in 1997 studied the killing habits of 1,000 cats. In five months they accounted for the deaths of 14,000 mammals, birds, reptiles and amphibians. With the UK population of 9 million pet cats this extrapolates to an annual toll of 250 million wildlife lives lost. Mice and voles form the bulk of the sorry corpses, with rats hardly featuring – so much for pest control! The studied cats were mainly belled, fat, middle-class moggies not the 800,000 feral mercenaries that roam with unmeasured ferocity. The bells appear to help reduce mammal but not bird deaths but the new sonic beepers such as Cat Alert seem to reduce predation on birds by up to 65 per cent. However, simply keeping cats in at night reduces their killing capacity by 80 per

*Top right: **Domestic Cat** with rat in mouth. The culprit caught in a most evil act of crime against wildlife. It's just lucky that I'm not the judge as I only have black hats!*

*Right: Making a run for it. A **House Mouse** dashes for the door and will live its life in constant danger of discovery. Sadly, in most instances, just a squeak is enough to set the cogs of destruction in motion.*

Right: A Domestic Cat on the prowl. Sadly, this purrfect predator, seen here, is squinting smugly at its unparalleled success as a suburban, SAS-style super soldier hell-bent on the war against wildlife. Look at those whiskers. Damn it, even I'm envious!

DO THE DECENT THING!

Lots of people 'love' cats and rightly so. They make good pets, are generally good with children and provide companions and comfort to the elderly. A little responsibility along with the 'love' can go a long way, and the first step is the chop. Yes, neutering. I'm told it doesn't hurt but it certainly stops a lot of pain – the pain of a dead bird a day. Please consider it. Phone around, some schemes charge minimally. Otherwise shell out and save poor souls.

cent and also cuts the daytime murder rate as well. There are 38 times as many cats as foxes in the UK and six times more than the sum total of all our wild terrestrial predators combined. Uncontrolled by their free food supply and pampered by their owners, the UK's cats are having a riot while the Mink and the Magpie have become hated scapegoats. It is clear that the real slaughter-masters are our cats.

Shortly before the publication of *Back Garden Nature Reserve*, where I penned a similarly poisonous passage about pussies, a real cat flap blew up over comments I had made about shooting cats. Alas, all was exaggerated, and the only weapons employed were water pistols. Nevertheless, the press conducted polls on the popularity of cats amongst the British public. Imagine my smugness when I recount that both Radio 4 listeners and ITV viewers phoned, wrote and e-mailed their reactions three to one against cats! The RSPCA acknowledged water pistols were an acceptable deterrent and launched a neutering scheme to reduce future populations of unwanted kittens. Naturalists countrywide came out of their cupboards to champion the cat-as-nuisance cause. At no time was cruelty advocated, but airing this increasingly relevant urban conservation issue can only be productive even if it is still thought to be provocative.

Pampered predators

Probably the highest density of feral cat colonies exists in the West End of London. The reason for this aggregation is linked to availability of food and necessary shelter. Here the human population is surprisingly high compared with most other sections of inner cities, which are almost deserted when daytime workers have commuted to the suburbs. There is also a high percentage of elderly people whose compassion ensures a greater number of dedicated feeders who freely dispense saucers of choice nosh. In these favoured city ghettos, security is enhanced by the large number of 'railings above basement' type houses which offer protection from the odd marauding dog and the feet of the daytime populous. The diet of these top cats is generously composed of over 75 per cent pensioner tidbits, supplemented by 20 per cent refuse scavenging, and rarely contains any natural prey such as mice, rats or birds. This is a refreshing result. Another interesting aside is that even in these inner city colonies, up to 10 per cent of a cat's diet consists of vegetable material, a curiosity for a carnivorous animal.

The lack of human delineation such as fenced properties and ordered roads in city centres leads to strong colony association – lots of cats living under the same roof. In contrast, the clear delineation and fenced human territoriality of suburbia provide individual cats with space and strategic points such as scratching posts to create and maintain functional territories of their own. Here each territorial tom returns to a separate base, typically your kitchen or fireside.

When sweet little 'Tiddles', 'Cuddles' and 'Muddles' stray into the urban jungle they became superlative predators. Their skeleton is constructed for speed. The front legs are hardly used for rapid movement: all the energy is generated by the bowed spine and back-legged thrust. Their claws are retractable and they are digitigrade, running up on their toes, quite unlike we plantigrade sloths. Their eyes are forward facing and adapted for nocturnal vision, being able to suck in 50 per cent more light than our own. They also have a tapetum, that shiny mirror which bounces back green flashes from your headlights as you cruise through the city at night. This structure

recovers any light which has strayed through the sensitive cells of the reflectors and bounces it back, enabling cats to see in one sixth of the light levels that we can. Having vertically slit-shaped rather than circular irises allows better control of light levels. Cats' hearing is better than that of dogs: their separation of closely spaced sound sources, their calibration of sound depth and their ability to hear the high-pitched squeals of their prey place them far ahead in the predator stakes. They are equipped with a sensitive bouquet of whiskers – strong thick hairs with enriched nervous connections. The 'moustachials' on their chin, the 'genials' on their cheeks and the 'superciliaries' fanning over their eyebrows are sensitive to air currents and anything they touch. Finally, their tongue is coated with backward-pointing spines, for rasping at their food as well as grooming their thick fur.

Cats are remarkable animals, wonderfully evolved predators that also make easy pets, the latter probably more pertinent in an age when they were also welcomed agents of 'pest' control. Mice and particularly rats were presumed to be the typical prey of cats, a fallacy I'm afraid, and these days no longer a viable excuse for keeping a moggy.

Indeed, rats are in decline, which is a shame because I feel that it is time that we recognized the rat as the single most important animate factor influencing the social, intellectual and artistic development of the western world. Yes, the rat, that beaten, trapped, poisoned and hated little rodent, has actually twisted the course of civilization and the way we all exist. Surely we owe it some acknowledgement?

*Above: Black and beautiful, but maybe in my eyes only. Once common, the **Black Rat** is now a national rarity and to my mind, its neglect in conservation terms is a national disgrace. It seems that even biologists can't forgive and forget its past links with plague.*

*Below: Dicing with death: an urban **Brown Rat** shows itself in daylight, probably on the lookout for bird food. If you don't want rodents to ramble around your garden, put the food out of their reach in feeders or on a table and store it securely, too. Remember, no food equals no rats.*

The Bubonic Burden

Great Britain is home to two species of rat – the Black and the Brown – both are invaders. When the Crusades came to an end in the Thirteenth Century, new trade routes spread across Asia and Europe, the coasts brimming with thriving ports importing exciting and exotic goods... and Black Rats. These stowaways carried a colleague, the flea *Xenopsylla cheopis*, which in turn brought pneumonic plague bacteria *Yersinia pestis*. Awesomely destructive, this disease, in humans, causes the glands in the neck, armpit and groin to swell into bursting buboes while a high fever produces red-rimmed eyes and tattoos black blotches all over the body. If the infection spreads to the lungs, pneumonic plague develops and a little later sufferers succumb to severe headaches. Death follows rapidly, usually within less than five days. If it is any consolation, the rats perish too, an inconvenience which leaves the fleas hungry for a bloody last human supper before they, in turn, die.

This chain of fatality is unusual because with each of the carriers dying, the possibility of the disease being effectively stopped in its tracks appears a distinct reality. However, the bubonic plague's remarkable virulence means that this is rarely, if ever, realized. Between 1346 and 1349, with the world hardly at the peak of hygienic awareness, one third of Europe's population perished, a minimum of 25 million people, meaning that this pestilence had killed more than all wars have ever done. The 'Black Death' continued to erupt across the globe until the Eighteenth Century and cities were always its most effective habitat. Close living permitted easy contamination and during the Great Plague of 1665, 98 thousand Londoners died at the rate of 8,000 per week. All things considered, I suppose it is no wonder that rats are not top of our zoological pops but how about forgive and forget? Plague has all but vanished as a threat to our health and rats are not what they used to be.

In fact, the Black Rat is extremely rare in the UK and has been absent from the mainland for the last hundred years, tenuously surviving on Lundy and the Shiant Isles in the Hebrides. It was an immigrant that arrived some time between 1100 and 1200. It prospered and because it was the plague carrier we have never forgiven it. Today, if you visit the Mammal Society's website you will find no fact-sheets, nor any mention of it on pages that consider

the UK's critically endangered mammals. The funky otter, the cuddly Dormouse, the bashful bat and all the rest – but there's no room for 'ratty'. Even dedicated mammologists and their students have no heart for the poor old rat. In June of 2001, a small colony of Black Rats was discovered in Cornwall. Did the conservationists rush to welcome this last chance to save the British Black Rat or to declare the kitchens where they were hiding National Nature Reserves? No. Pest controller Spencer Rickard killed them and said that after twenty years of rat bashing he was pleased to have finally seen the real Black Rat. I bet the rats weren't as pleased to see him! To add insult to fatality the BBC released a load of news about the plague and advised people not to panic. Brilliant – xenophobes, rodenticidal maniacs and an establishment obsessed with the cute and the cuddly. I have never seen a Black Rat in the wild and, given the above, I doubt I ever will. That's sad. I think they're very attractive and thankfully I am not completely alone.

CYBER-RATS

Clare Jordan runs an excellent Black Rat website dubiously entitled 'Hello Sailor' where all the ratty-facty-info you need is enthusiastically displayed. She tells of the Brantons' attempts to domesticate the species using, ironically, rats procured from Rentokil. This couple learned a lot but they discovered that you cannot make a pet from this ex-purveyor of pestilence. Black Rats are too skitty, are parrot-like in their partiality to particular people, can be unbelievably fast and bloody-minded, and they bite. One in a million will love you to death… well not literally, not any more. Oh, forget it!

Well, not quite yet. I checked out Brown Rat on 'the net' and found a lot of 'rattist' nonsense: this species is fierce, aggressive, will kill goats, lambs and calves and when you are in London you are never more than 5m from a Brown Rat. What a fantastic invention the Internet is! I am going to burn all my books, take my brain out and log on to mis-information dot com for ever! Never more

*Above: Two **Brown Rats** overheard whilst home-hunting. 'Well dear, the outside seems sound enough!' 'Yes, but the interior is a bit grubby. Do you think we could get the Changing Rooms crew in to help?' 'I think so, especially as we're living in a water feature.'*

than 5m from a rat! One might presume that the sewers of the city are a seething refuge for rodents. Sadly, they are not. The modern sewer is clean, rat-free and, besides, the number of rats is related to the amount of rubbish available, and the amount of rubbish is related to the wasteful or slovenly folk who drop it everywhere. So who are the real rats?

STREET LIFE

To the average precinct pedestrian, a flock of Feral Pigeons appears as a swirling mass of tatty or scruffy birds. But, in fact, this flock is a highly organized group where strict hierarchies govern the status of all the birds. The dominant pigeons strut about in the bustling centre where they find more food per minute over a smaller area. The subordinates are relegated to the edges, find less food and are consequently lighter and less fit, and for city pigeons, food equals fitness. In summer, they spend up to 50 per cent of the daylight hours pecking at crumbs – mainly early in the morning and late in the afternoon – but in winter, when their metabolic demands increase, up to 90 per cent of a pigeon's day is spent just searching for food. Within the city, this includes almost anything edible: bread, biscuits, cake, peanuts, banana, apple, potatoes, cheese, fish, fat, meat, chocolate and ice-cream in any combination and in any state of decay. They will also eat invertebrates and thus worms, snails, small insects and spiders all provide tasty treats.

Having such a cosmopolitan palate is key to the Feral Pigeon's success story.

Perhaps, surprisingly, given the feathery chaos on the pavement, pigeons are monogamous and pair for life. Some couples remain together all year round, continually doting on each other and often mutually preening. This apparently affectionate act may serve to remove parasites, but more importantly, it is a signal, a form of brush-off, as it indicates that a female is not ready to mate. Such rituals of street pigeon behaviour have been thoroughly investigated, are simply explained and easy to observe during any city centre lunch break. First, pick your seat, second, open your sandwiches and third, prepare for the attention of the starving hordes. Once you have set your bait, start watching closely.

'Driving' is a behaviour often seen in groups of foraging pigeons. A male will strut deliberately on long rapid strides behind a female, matching her every turn, sometimes even treading on her tail in an effort to keep up. His aim is simply to drive her away from the amorous attentions of other males now that she is receptive to their sexual attentions. Mating may follow such a chase, with the female suddenly stopping and begging the male by gently pecking the base of his bill. This kiss soon becomes more passionate, their beaks entwine and both perform regurgitating motions. This gesture is largely symbolic because only a small amount of food is passed between the lovers. Next the pair circle one another, the male mock preening by thrusting his head backwards into his shoulder feathers momentarily before mating. The female stoops down, raises her wings slightly and cocks her tail, and while she thinks of England, the male hooks his leg over her tail and in a flapping frenzy attempts to copulate. It is not a very erotic affair, nor even very successful judging by the number of attempts made, but then, lacking any intromitant organs, most birds rely on aligning orifices to effect insemination which is pretty difficult. However, mating accomplished, or at least attempted, birds

Below: Feral Pigeons.
Feed the birds, tuppence a bag. Not anymore! Not in Trafalgar Square, thank you. The bird-food seller's license has been rescinded, so if you want to make thousands happy for next to nothing, you have to bring your own. Please don't forget, the hordes are still hungry.

now parade about, heads held high, wing feathers drooping and rump and ruff feathers held erect. They high-step across the pavement brimming with confidence, and often the male will follow up by performing a display flight. This is characterized by slow and exaggerated wingbeats, followed by several loud wing claps and a rocking glide down on V-shaped wings. It serves not only to signify a successful sexual encounter but also to mark a territory and to attract a mate in the first place.

Muggers and murderers

Aggressive behaviour in pigeons is slightly more complex, although many of the antagonistic and submissive postures or actions double up as sexual displays. Wing twitching, where one or both wings are repeatedly flinched upwards, precedes a wing cuffing bout, yet these actions are often performed by both male and female after successfully mating. The very obvious 'bowing' display, where birds erect their neck, crown and back feathers into an upright position and then vibrate their swollen chest as they gently bow, serves both as a male-to-male confrontation and as a sexual attractant to a mate. In the latter instance, the behaviour is slightly modified as the tail is fanned, closed and depressed as the nodding, spinning and strutting male dances round the female.

One behaviour not sympathetic to both causes is fighting, and pigeon fights can be vicious, even fatal affairs. Opponents deal powerful and rapid blows, using the joint of their wings to pummel the other's head. As the bout escalates,

these rivals peck at each other's face, attempt to grip each other's bill, and shake and tread over each other. Feathers fly and blood spots the pavement until at the end of the contest the weaker bird retreats injured and exhausted, while the victor parades about bowing.

Far from those overindulged and over televized African plains, the city centre precinct is a comparable hotbed of lust, love, loathing and violence. It is a 'soap' replete with all the nastiness, rivalries, controversy and characters that seem to typify that television genre; except it is real, and for me at least, far more interesting.

SKYSCRAPERS: BIRDS

The best things in life are birds and the urban fauna contains some of the best birds on offer. If you haven't got your binoculars with you on the bus, you will miss out on some special views, not necessarily of national rarities, but of some confident common species that provide us with the best opportunities for close-up study of their behaviour and ecology.

FROM RUINS TO ROOFTOPS

In the summer of 1975, I took a train to South Wales and thus a timid thirteen-year-old turned out onto a Pembrokeshire platform determined to do his bit for the best bird in Britain, an enigmatic, rare and remote raptor that we all believed was doomed to extinction. The Hawk Trust provided the caravan, Wales provided the weather, my mum the warm clothes and a pair of Peregrine Falcons one of the best weeks of my life. I wasn't wet and windblown, undernourished and unpaid, I was in ornithological nirvana and like so many other folk of all ages I fell in love with these fantastic birds. I would have died for them – thinking back, I probably nearly did I was so cold! What a contrast in that now, 25 years later, I can be shopping in my home town of Southampton and look up from the precinct and see a Peregrine circling. Now I can sit on the embankment in the centre of London and watch one fly between the buildings or enjoy the same thrill from the promenade in Brighton. The Peregrine has escaped extinction to make a sensational comeback and now it's coming to town.

Peregrines have always had a high profile, prized by princes for falconry or persecuted by the same to preserve game, they've been forced to fit with the fashion and have thus enjoyed mixed fortunes. In the 1950s they found themselves flying fast into the face of adversity because despite having survived 200 years of genocidal gamekeepers they met their miserable match in organochlorine pesticides such as DDT. The pesticides were used as seed dressing and accumulated in the bodies of the Peregrine's seed-eating prey. The Peregrines, topped up with the toxic metabolites of these chemicals, paid the price for being at the top of the food chain. Adults died or began producing eggs with shells so thin they were broken during incubation. Breeding success became widespread failure and Peregrine numbers fell from an estimated 1,100 pairs in the 1930s to only one hundred successful pairs in 1963. All too belatedly the pesticides were withdrawn and the remnant population of these raptors began a slow recovery, assisted by thousands of volunteers like the young Packham and confounded by the lunatic interest of egg-collectors and unscrupulous falconers. Today we have about 1,500 pairs of Peregrines and we no longer have to go to wildest Wales to see them because, in gratitude, they have come to us.

This success in our towns and cities came rather later than in the USA. The Department of Environmental Protection has run a Peregrine

Left: Peregrines are in the Top Ten of anyone's bird list. Potentially the fastest animals on earth, stooping on their prey at speeds in excess of 100mph, their dashing lifestyle is awe-inspiring. The fact that all our cities provide an adequate resource for these cliff-nesting, pigeon-pinching princes is a remarkable bonus.

Project in New York City for the last 15 years, ringing, erecting nestboxes and studying prey selection. There are 14 breeding pairs in the city, the greatest concentration of urban Peregrines anywhere in the world, and whilst the majority of eyries are on bridges, others command spectacular views from Manhattan's downtown office blocks. The Hawk and Owl Trust has been monitoring urban sites across the UK since 1998. It is slowly uncovering the complete picture but is often hampered by the sincere secrecy with which people still shroud anything 'Peregrinny' in, despite the fact that city breeding sites are probably the safest countrywide. There is little chance of finding poisoned bait on the pavement,

Right: Kestrel. The urban living of this pretty little falcon was first brought to my attention by the Blue Peter team in the 1960s, when their idea of a novel nest was a window box in the city. These days, however, no tower block should be without one.

Right: What a sneer! A Peregrine giving us the ultimate put-down. Self-assured and superb – if only he played for England (or Scotland or Wales or Ireland). Then we'd be rolling in World Cups, Wimbledons and wickets won. Confidence is the key, and in this killer it is plain to see.

days and it's better to be safe than sorry when it concerns something as precious as Peregrines.

In the summer of 2001, a total of 64 UK breeding pairs were using man-made structures for their nests – a significant increase from the seven pairs making a similar choice in 1991. Of course, not all of these birds were urban or city-based but amongst them appeared the first confirmed breeding for the capital, three young birds fledged from an eyrie on Battersea Power Station. This is an auspicious edifice indeed but perhaps not so revered as Exeter Cathedral upon whose architectural splendour another pair have procreated. Some other sites are equally well known and publicized: the Brighton birds always make the local news thanks to the ingenuity of Graham Roberts of the Sussex Ornithological Society, who has fitted CCTV cameras inside the nest box they use, and learned a lot about 'them upstairs'. Typical for raptors, not all of these sites are used for breeding every year but there emerged a number of other year-round roosting sites that were in continual use and where we can hope that breeding will occur soon. Thus Basingstoke, Bath, Glasgow, Gloucester and Manchester seem set to join the lucky list in the not-too-distant future.

There appears to be a definite correlation between these city colonists and water, whether it be lake, loch, river or estuary. It seems that a diet of pigeons and Starlings is not enough and that waders, wildfowl and a huge range of migrants swell the Peregrine's menu into the most cosmopolitan of its kind. At the top of the Brighton block, on a ringing visit, I discovered the remains of Sandwich Tern, Woodcock and Water Rail amongst a fluffy

shotguns on the balcony or egg collectors with very long ladders. Of course, all this cloak and dagger stuff can be frustrating but the Trust maintains a strict confidentiality; after all it is early

Left: *This tit has been killed whilst still inside the squirrel proofing on a bird feeder. This says a lot for the dexterity, determination and guile of the **Sparrowhawk** who now has to work out how to remove its prey from the cage. Peregrines may be fast, but they're not as furious as these little predators.*

flotsam of more predictable prey. Another trend that this research is beginning to suggest is that birds reared on bridges go on to breed on bridges, those hatched on pylons, or churches, or tower blocks mature to choose eyries in or on the same. Although such sites might be secure, they are not always the most productive: exposed flat roofs invite chills for the chicks and tall, linear and ledge-less buildings are not great for fledging from. If the youngster's first flight ends in a grounding it is unlikely that adults will feed it, particularly in street situations. Thus the more suitable urban eyries are those where nearby structures closely replicate a natural cliffside with all of its ledges and crannies and with supportive currents of uplifting air. So, if you ever find a Peregrine in your parking spot, please don't presume that it is tough enough for street life because in the city at least, the sky is its limit.

Problem Peregrines

While raptor fans will relish the increasing number of urban Peregrines, there is one group who feel radically less enthusiastic about the prospect of speedy stoops over the rooftops. Pigeon fanciers first lobbied the Government to control Peregrines in 1925

Left: *In recent years **Buzzards** have spread eastwards from their stronghold in Wales and the West Country. Whilst their habits make them unlikely city dwellers they do land nearby when scavenging road kill. As our cars continue to produce carrion, these birds will continue to prosper on the routes in and out of town.*

and today their protests are heard again. During the Second World War, about 600 Peregrines were culled in the south of England in an attempt to maximize the success of homing pigeons

*Right: From woodland to wasteland, the **Tawny Owl** is the only owl species that has successfully come to town. A diet of birds instead of mammals and the security of mature parkland trees for roosting and nesting are what it requires.*

*Right: This is not the result of a crash landing, this is a **Mute Swan** enjoying a good bath. Although our large park ponds always seem to have their resident swans, they are a species in national decline. Try to vary the diet you offer – get some maize or corn – after all would you like stale bread every day?*

carrying military messages from the Continent. Local extinctions resulted but such impacts have been long rectified, and thus, in 1995, the pigeons' pals put pen to paper again and in response the Government commissioned a thorough investigation by the Raptor Working Group. Under its auspices The Hawk and Owl Trust put in three years of rigorous research and analysis, and the result, 'A Study into the Raptor Predation of Domestic Pigeons', was published in March 2000.

Quite alarmingly this revealed that just over 50 per cent of racing pigeons fail to return home, 20 per cent because they stray from the race route, 20 per cent because they collide with overhead wires and 8 per cent because they are shot, oiled or poisoned. Only 3.5 per cent are perceived to succumb to Peregrine predation and most of these would have already reverted to a feral existence or would have strayed great distances from the race route before they were killed. Such losses were highest to Scottish and Irish lofts and significantly lower

Above: *A* **Starling** *in silhouette. A couple of streetlights. Drive by. Ignore. No, look. A bird. It was common, not now though. Here today, gone tomorrow, maybe. It's a mystery. We didn't care for them. Noisy bullies. 'Pests', we said. Killed them we did. Poor Starlings. Telegramese complacency? Let's. Hope. It doesn't. Come. True. Beware. Stop.*

Right: *Splatter pattern. The imprint in feather dust left by a* **Collared Dove** *on a pane of glass. Big windows reflecting the garden are a problem and many birds are killed by such impacts. Either draw your curtains when you're out or put sticky markers – spots or bird silhouettes – on the glass.*

in eastern England, a reflection of Peregrine distribution. Pigeons were killed irrespective of their colour, age, sex or racing ability.

Faced with such apparently insignificant figures it would be easy for us non-pigeon fanciers to dismiss this debate out of hand but racing pigeons is surprisingly popular with over 70,000 people keeping 3.8 million birds in lofts across the UK. In August, on the busiest day of the racing season, almost a million racers are released all over the UK and from France and Spain as well. Some birds are loved, others cherished, some almost worshipped and most keen keepers are truly devoted to their bird's health and welfare. There are also pricey pigeons from super-stud-stock but this sport is more about passion than pence and this is reflected in the fierce reaction of fanciers who believe their losses are due to Peregrines. In some cases the guilty verdict is clear-cut as the racing birds' rings are found in or beneath the Peregrine eyries.

Clearly, there is need for a compromise here but in my opinion not one that involves the legalized killing of Peregrines. These birds have become a conservation icon. The public have been repeatedly told that the Peregrines are precious and for that reason alone we cannot start shooting them. It would be too confusing: one minute they are sacred, the next they're a pest. Who is going to put a pound in a pot to help the next bird in bother if the last ones have been blasted? It seems we should find out why so many pigeons stray to reduce losses all round. Postpone the 'old bird' racing season by two weeks to shift it out of the raptors' breeding season, route more races through the east of England where losses are less, and more thoroughly investigate the visual 'eye' and 'target' deterrents which some fanciers attach to their prize pigeons. If this sounds like pigeons having to fit around Peregrines then so be it; after all, how many teenagers would sleep in sweaty tents in the middle of freezing nowhere to protect a pigeon? In this world where all birds are equal, some are definitely more equal than others.

*Left: Clever bird the **Carrion Crow**, maybe cunning too, and definitely conniving. It is cautious and watchful but keen to seize any opportunity to prosper. Whilst it is not a common city dweller, it can often be seen on the prowl for picnic leftovers or picking amongst the flotsam at low tide.*

RETURN OF THE RAVEN

*Above: The scapegoat. 'If there's trouble in the garden then blame the **Magpie**', is the imprecise policy adopted by many. Yes, they've increased in number and spread, and yes, they are obvious but they are not guilty of song bird genocide – McCavity and his mob of 'moggies' have blood all over their paws!*

A bird that has no equal when it comes to a reputation for murder, mischief and malevolence is the Raven. Odin's advisors, Poe's conscience, Ravens also represent for many people a dark and ancient wilderness, but for all their wild and windswept legend this is a recent affectation. Two centuries of persecution drove our Ravens to the remote north-west coast of Britain. The pesticide crisis then nearly pushed them over its edge and so it was here that we who searched saw them tumbling and kronking in the tempest and Atlantic spume. But in the days when our attitudes were lenient and our streets full of rubbish, the Raven was a quintessential part of any city's bird fauna. Those famous squatters at the Tower of London feasted on the squalid contents of the castle's moat, which for hundreds of years slowly filled with waste and although only six sit, clipped-winged, on the ramparts today, once sooty clouds swirled above the executions

*Below: The voice of menace is uttered harsh and low, but it seems that rather than 'Nevermore' the **Raven** might just be 'Everywhere' if the current trend of range expansion continues. If only we were more messy, more unhygienic, more medieval, then they'd have a riot.*

Above: Jackdaws need to be social and need nesting sites that provide for colonies rather than single pairs. They are hole nesters, so buildings will do the job, preferably derelict buildings which are close to diverse foraging opportunities, because just like all the best urban champions, they can eat anything!

Right: The stunning result of perhaps the most fantastic bird conservation success story. The re-introduction of the Red Kite has given real hope that both this and other raptors will not only survive the Twenty-first Century but restore their populations to nostalgic highs.

and blood-lusting crowds. Just like the Peregrine, Ravens have been bouncing back of late and those that have chosen to nest on Chester and Liverpool Cathedrals have become national celebrities. Winter roosts continue to swell, particularly in the Midlands, and it surely won't be long before fully feathered Ravens will be rolling over most of our cities, particularly those cities on rivers or the coast, the shores of which provide them with rich pickings in our otherwise too tidy times.

MIGHT THE KITE?

Following in the Raven's wingbeats, and sure to cause great excitement, is the splendid Red Kite. Until recently, a tiny population of inbred birds eked a meagre existence in central Wales, but following a highly successful series of countrywide re-introductions, these stunning raptors are spreading all over Britain. Sadly, the flocks that once wrestled with Ravens for streetside rubbish are unlikely to return due to unsympathetic standards of hygiene and refuse disposal. But urban fringes should enjoy patronage and, if the legions of urban bird feeders were to include a little meat amongst

their otherwise seedy offerings, perhaps the impossible might happen. Indeed, as I write, a magazine has appeared on my desk, which excitedly reports a Red Kite having been seen circling over our capital city. Not that London isn't often good for spying a national rarity; in fact it has got one all of its own.

BUILDING FOR THE URBAN BIRD

Upright, twitchy and quick, in a dusty black coat with a brick red disappearing tail, the male Black Redstart is a great little bird and, above all, in Britain at least, it is the definitive urban species. That is to say, it doesn't occur in any of its primary natural habitats, only in our cities. A Robin-related member of the thrush family, it struggles here on the northern edge of its range, more at home on the warmer, insect-rich rocky slopes and screes of southern Europe. However, postwar bombsites, old docks and redundant industrial land physically replicate its favourite hunting and nesting habitats, and the extra few degrees of temperature above ambient in the city

Above: A **Robin** *redbreast in waiting. This young Robin will have to wait until its adult plumage appears in its first autumn before it gets rowdy. On first leaving the nest, the speckled nature of these feathers help whilst it is ground-bound.*

Left: On the northern edge of its range in Europe, the **Black Redstart** *might be one of those species set to increase with global warming. Already some insects have started the trend, and the next few years should reveal whether birds such as this will follow.*

*Above: As much a part of the park posse as the beach brigade, **Black-headed Gulls** have become accomplished picnic pinchers and large flocks are a permanent feature of city waterways. However, they return to the coast to breed, leaving only the young birds and celibates to scrap over your sandwiches.*

*Above: Top of the Pops in town is the **Blackbird**. Rich, varied, mournful and sweet, the song of this familiar species should never fail to delight, especially when delivered with such gusto.*

centres helps with winter warmth and keeps the insects moving longer and later. Despite our run of milder winters the population has fluctuated for the last 50 years and currently flits between 75 and 100 pairs, making it one of our rarest birds. Sadly, its future conservation will not be easy. Following the publication of the Urban White Paper in 2000, current development practices favour building on 'brownfield' rather than 'greenfield' sites, a policy that appears to suit most of our ideals of wildlife preservation. Also, because most of these areas which are Black Redstarts' breeding sites are deep within city centres, they represent priceless parcels of prime real estate. So is there a compromise that may give conservation a chance? Yes, and its champion is Mr D Gedge.

Dusty Gedge is the lead for the Black Redstart Action Plan in London and regularly acts as a consultant to property developers, architects and various agencies. He is not one of the old school of old-fashioned single species conservationists; he has bigger ideas, namely 'Brownfield Biodiversity' and his ambitions are at the highest level, literally, because Dusty's solutions are on the roof.

On the continent, 'green' roofs are common place and often a statutory requirement in Germany and Switzerland. In Britain, as ever, we've been too conservative in our approach and aside from the top of the odd environmental centre you're unlikely ever to encounter the phenomenon. Some such roofs are genuinely green, planted with grass or sedum, but many are essentially roofs which have a substrate added to act as a sponge, typically crushed brick or concrete. Of course, such a surface is rapidly

colonized by the local ruderals but often sparsely and thus the green roofs are in fact grey, which is not a handicap because this closely resembles the redstart's brownfield requirements. In Europe Black Redstarts or Biodiversity are not the most important issues for planners considering green roofs. Reduction of water run off as a flood protection is top of their agenda, followed by energy conservation through improved insulation and access to visual green space, which sees many green-roofed hospitals given the go-ahead. Dusty's drive to encourage the practice here has been frustrating as investors, planners and buyers are simply unfamiliar with the continental habit, and even other conservationists have displayed scant regard for his ideas. However, his perseverance is paying off; several schemes in the capital will be putting grass on a different level. Depford Creek and Gargoyle Wharf will see extensive green roofs for the benefit of redstarts and others and, in fact, green roofing systems are going up where these illusive birds have never stretched their wings. Dusty is also dealing with architects and developers at an earlier stage as they become keen to incorporate eco-friendly features into the initial design and thus save costly changes at a later stage, for planners too are now recognizing brownfield ecological issues. English Nature is currently investigating the role green roofs can play in urban biodiversity and as more go up, and familiarity spreads, Dusty hopes that the roofs will cover urban Britain in time to secure a safe future for his favourite Black Redstarts.

SPARROWS ON THE SLIDE

You may wonder why stars twinkle, or why footballers get paid such ridiculously high wages or where all the House Sparrows that twittered around your eaves, chirped in your dust or scowled from the top of your shed have gone? The first can be explained but read it elsewhere, the second is a complete mystery and the last is grave cause for concern, so grave that a national newspaper has offered a £5,000 reward for a satisfactory answer.

Once so common they were decreed a 'pest' by farmers and persecuted accordingly, so common that my mother only gave them bread on the bird table, so common that we all took these cheeky little granivores for granted, House Sparrows are now seriously declining, and fast. The British Trust for Ornithology's Common Bird Census has revealed a 58 per cent collapse in numbers in rural Britain in the last 25 years and a 53 per cent decline in urban areas in the last five years. Sparrow expert Dennis Summers-Smith has charted a 96 per cent demise in the urban centres of London, Edinburgh, Glasgow and Dublin and I haven't seen one in my mum's garden for about two years. Was it the bread? No, it was not, but precisely what has instigated this woeful demise is unknown.

Summers-Smith has the theory that the toxic compounds that have replaced lead in petrol have drastically reduced insect densities in our urban centres. This has had a

*Left: Like Robins, **Wrens** can adapt to our ways of doing things, and despite their typically fidgety and skittery nature, individuals can become equally as tame. Winter-feeding is the key to attracting them. Use live food – mealworms or maggots are top of the menu.*

Left: *Once a sight so familiar,* **House Sparrows** *were universally ignored. Now they are both snapped and studied to plan their future conservation. You can help. Put out grain and put up nest boxes under your eaves, especially if your roof space is otherwise impenetrable.*

knock-on effect as insect food is essential for the early development of nestling sparrows. Radical maybe, but along with over-tidy, over-pesticided gardens, modern houses with no nesting sites and a countryside that's been run with the 'not a crop – not wanted' mentality it is not surprising that this once successful species has taken a very sad plunge. I've no doubt that a complex of ill conditions is to blame but feel that the harsh intolerance of modern farming and its ruthless efficiency lie at the heart of the matter. Herbicides and pesticides ensure weed-

Left: *It's this worm's turn to nourish the thrush family. The* **Song Thrush** *is doing well in our gardens, but disastrously in the countryside – a familiar story for a host of once familiar species. If you lay off the slug pellets, you'll be doing your bit.*

free crops and kill all the insects, no grain is spilt after harvesting, no stubble is left, and worse, harvesting occurs earlier in the year, leaving no food at the end of the breeding season for the young 'spadgers'. Strictly speaking, the House Sparrow is a seed-eater and it is likely that rural populations would always have continually supplemented our city birds in their marginal habitat. (I'm sorry, Mum, corn would have been better than bread so you are partly to blame, and while I'm at it I must say that you should have taken greater care of the Starlings too.)

GONE FROM THE GUTTERS

These summer-spangled, winter-spotted wonders have decreased by 65 per cent since Armstrong trod on the moon. They are likely to be upgraded to 'red status' in terms of being a bird of conservation concern. It is insane, the Starling, a rarity! How did it all go wrong? Well, the finger again points at intensive farming practices and it is such a shame because if ever there were a truly 'splendid' bird then it is the Starling.

These bawdy ruffians were the first avians to make my acquaintance as my mother pushed my pram beneath the apple tree in our garden and I marvelled at the 'dirdies' squawking above me. My father suggests that their falling faeces infected my interest in all things feathered but such seeds were more likely sown when, at the age of five or six, I picked a dead Starling from the gutter by the bus-stop. Never had I seen such beauty, the neat rows of beautiful glossy feathers all so perfectly designed. The rich petrol sheen of the bird's wing coverts were a marvel to me. Little did I know that no pigment was involved. It is the interaction of light passing through tiny chambers in the feathers' barbules, those microscopic comb-like structures that bind the feathers together. The reason that only the coverts and not the primary flight feathers are iridescent is that to glimmer effectively the barbules must lose their hooks and the feather's structure is significantly weakened. Much later, on a sunny afternoon, I discovered on the grass outside our house four perfectly smooth and unbroken Starling eggs. I presumed that the female had for some reason got 'caught short' and laid them there on the ground. Of course, this was juvenile naïvety and it has since been discovered that this relatively familiar habit is the result of eggs being removed from the nest and deliberately placed outside. Furthermore, it is likely to be something to do with this bird's habit of practising incipient parasitism. Some females deposit their eggs in others' nests without ever building their own and, as a consequence, a large percentage of young Starlings are not at all related to either of their apparent 'parents'. This is surprising as Starlings exercise a strict monogamous bond during the breeding season. The males give particular attention to their females to prevent

*Opposite: Safety in numbers. Like all regular performances some nights are better than others. A breath of wind to get them swirling, a sunset to turn the sky orange and a late bulk arrival of the flock to ensure a cast of thousands – these are the ingredients required to produce the best of the **Starling** ballets.*

*Far right: Sick of kids screaming, sick of feeding grubs to the wife, sick of parasites and dodging predators: this looks like a **Starling** who needs to flock off to the countryside with his mates, on a weekend of rural rioting, raiding cherry orchards and pinching grain.*

*Right: The **Starling** is an excellent mimic, copying other bird calls – one in our street does a brilliant Tawny Owl – and even man-made noises. I'm not sure whether the oft' quoted references to its referee's whistle are true, but door chimes and reversing lorries I've heard for myself.*

*Right: This piece of litter has been jewelled by a **Kingfisher**, a species that needs clean and clear water for its plunge-fishing habit to prosper. Minnow, Stickleback, Gudgeon or fry fit the menu, so urban visits are not uncommon out of breeding season.*

any of the excess of suitors from nipping in to sow their genes amongst the nest holders.

The Starling's beauty is not confined to its plumage. In the recent past huge numbers of migrant Russian Starlings would arrive each winter and roosts of millions once occurred in the forestry plantations of East Anglia. Indeed, many of our city centre parks also played host to these astonishing gatherings. Unfortunately, concern for hygiene prevailed and the birds were shot, gassed, poisoned or pushed off by having lights shone at them or tapes of predators played to them. In some extreme cases the roosting trees were felled to prevent the birds from aggregating. In the late

BUILD A BOX

Typically we are fond of erecting bird boxes in our gardens, particularly tit boxes which are available off the shelf at garden centres, but why not put up some boxes at work? The type of box chosen should reflect the range of species you may feasibly attract, everything from Starlings to Tawny Owls. Give up a Sunday morning to saw something together in the shed and have a word with the boss. With a little effort and perseverance you could make a pair of tits very comfortable, and there's nothing wrong with that.

1970s and early 1980s, as the sun set on London's Leicester Square, great numbers of Starlings would gather and for half an hour they would raise a cacophony and command the attention they deserved. To the best of my knowledge Starlings are capable of spreading a little gastro-enteritis, they may pass on tape worms, a few protozoan parasites and some fungus but I suspect that what people really didn't like was the mess they made on the ground below. If only we had known that the bird was set for such a rapid demise perhaps we would have been a little kinder to those winter tourists. However, in a few places the Starling's greatest performance can still be seen. Brighton

Left: The **Grey Heron** *needs shallows to stalk its food – anything from fish to frogs – but will use canals such as this one as routes between its feeding or breeding sites. In towns and cities, islands on rivers are chosen for the bird's heronry as it is very conscious of access to predators such as man, foxes and rats.*

cannot fail to stir even the most cold-hearted council health officer. If I were you, I'd catch these performances as soon as possible because the future looks bleaker than Siberia for our old friend the Starling.

It would be a shame to conclude upon a passerine in peril, particularly when so many of our urban and garden birds have shown significant increases. In the last 25 years, numbers of Pied Wagtails have increased by 45 per cent, Carrion Crows by 73 per cent, Jackdaws by 80 per cent, Woodpigeons by 88 per cent, Magpies and Great Spotted Woodpeckers by 102 per cent, Sparrowhawks and Buzzards by 188 per cent and 350 per cent respectively, and Collared Doves by an astonishing 638 per cent. So it is not all bad news in the urban bird world even if the Magpie's increase is generally not so popular.

I think we have to be far more pragmatic in our perception of our unnatural world. It is not just our cities that are manscapes, every hectare of Britain has

and Aberystwyth Piers are well worth a visit in the colder days after Christmas, not to marvel at the Victorian architecture but at the swirling crowds of Starlings that mass each dusk to roost beneath these crumbling hulks. A beautiful, twisting cloud forms through a graceful ballet performed by a cast of thousands of speck-sized birds. The spectacle retains a primal charm that

been modified and for good or bad the results are bound to lead to changes. It would be easy to lament the loss of the House Sparrow and the Starling, but we shouldn't forget that the Sparrowhawk, the Red Kite, the Peregrine and a host of ex-woodland and scrubland refugees are perhaps more than compensating in terms of ornithological excitement.

KEEP YOUR EYES PEELED!

Even if raptors are beyond the range of your binoculars, the latter should always be in your handbag or briefcase, whether shopping or working. Many of the tiny, lightweight binoculars available are worth the price of purchase and trouble of carriage as the images they produce are excellent in good light. Times to be especially keen are April and September when the migrants are on the move and many follow natural or manmade features such as rivers, valleys or woodlands, or canals or railway lines. And so warblers such as Whitethroats, Blackcaps, Garden Warblers, Chiffchaffs and Willow Warblers work their way across country through the scrub and undergrowth. House Martins, Swallows and Swifts swoop above the water, and larger species fly higher over those features they use to navigate. At night, in rare moments of city

silence, wildfowl can be heard as can the piping notes of waders such as Redshank, Greenshank, Oystercatcher, Dunlin and sandpipers.

During winter, the berry-laden bushes so frequently planted to screen new roads, car parks or create borders in housing estates attract a range of thrushes, including Redwings and Fieldfares. They are also famous for feeding

Above: A Post-modernist Moorhen makes use of a suitably flamboyant backstop to place its nest safely in the centre of the stream, out of the reach of today's kids, foxes and cats – none of whom like wet feet.

Left: Look! An ice-cream van! You tell me what that common urban bird is and I'll get you a lolly. Herring Gull. Well done! Actually, I've only got enough for a mini-milk. Sorry.

Right: Some nuts and a rotten tree are all the **Greater Spotted Woodpecker** needs from us to get by. This bird deals with the insectivore part of the diet itself. It is so accomplished that it has increased fantastically of late and its drummings or laughing call can be heard in cities countrywide.

those sexy Scandinavians, the Waxwings, when, during cold winters small flocks grace us with their extraordinary beauty. These birds are relative rarities of course, but full-blown national, twitchable, super-stonkers occasionally turn up in cities. Long ago, I saw a Lesser Yellowlegs (an American wader) in a muddy puddle on an Eastbourne industrial estate. A Golden-winged Warbler alighted in a Safeway's car park and attracted more than 3,000 birders, and a Baillon's Crake found a home on a tiny, grotty park pond in Sunderland much to the delight of the twitching community.

All in all, if you really worked hard, and were constantly on the lookout for birds, all year for several years, in any major British city, there is no reason why you couldn't see more than one hundred different species. Of course, things begin to slow down and you will end up needing to see some pretty special birds to get a new one for your list, but there's nothing wrong with that! I'm afraid the chances of a Nightingale in Berkeley Square have probably passed for ever but then if you don't look you will never know!

THE GREEN TEAM: PLANT LIFE

On the micro-scale, you would be forgiven for thinking that the extremes of the man-made environment and its fabric are simply too hostile for the frailty of the plant world. But no. Plants are pioneers like no others, and the capacity of an astonishing diversity of species to germinate and make green the urban environment is phenomenal. Only our constant attention prevents our concrete jungle from rapidly reverting to its more primal namesake.

LEAVES AND LITTER

The shimmering glass and steel walls that rise to scrape our skies are an inhospitable environment but the concrete cliffsides that form the deep canyons criss-crossing our inner cities sprout sprigs of twinkling green. From the pavement to the sky these dirty gashes are shady and only when the sun is high overhead are they lit, yet plants still defiantly grow here, gesturing a feeble two stalks at our concrete blanket. Behind Covent Garden in London, a wreath of anaemic ivy sprouts from crumbled mortar high over the rabble of the opera crowd; above six high doorways near Manchester United's football ground creeps a prickly garland of bramble; in the central reservation of the M8 running through Glasgow a tiny colony of emaciated Silverweed survives in what must rank amongst the most harsh environments on Earth. On the micro-scale just about every national monument, famous shop, school, sports ground, church, factory and house in Britain is decorated with a small patch of moss or lichen. And it is these two plant types that more often than most find ideal conditions in the manscape, coating and encrusting the concrete and stone as they would the barren slopes of any rocky wilderness.

Mosses and lichens; no two groups of 'plants' are so ignored. I checked. There are no international lichen shows and moss fanciers are few and far between. Universally trodden over with only seaweed seen as a rival in tedium, their green carpets or crusty scabs are common in our cities but are regarded as boring with a capital 'B'. Nevertheless, it pleases me that they have not escaped the attentions of the world's most fastidious naturalists, even under the most unusual circumstances.

Right: A tower of ivy provides a fabulous resource for a wealth of wildlife. Far from being a danger to brick or bark, this much maligned plant can provide food and shelter like few others, and should be tolerated if not actively encouraged.

Right: Sphagnum Moss.
Mosses might be primitive but they are still pretty. Not in a gaudy, flashy, flowery way, but in a delicate, subtle, textural fashion. Their simplicity means that identifying individual species is difficult, but the range of habitats where mosses can prosper is surprising, from bogs to brickwork. I know they're not that sexy but life takes all sorts.

SPREAD YOUR SEEDS

It's childish I know, but I still take delight in scattering or placing seeds of the most stoical species amongst the places in the city I regularly visit. I do this, firstly to see just how tough these plants are, secondly, to give them a hand, and thirdly, and perversely, to fight against the horrible urge towards tidiness we often have. Thus, I have a Birch seedling, now tree, growing from behind a wall at Southampton University, a Buddleia and a dwindling bank of Red Campion in the city centre. It's simple. Collect some seeds, packet and pocket them and secretly sow them to add a splash of green to your own city.

SAVED BY MOSS

In 1799, the cavalier explorer Mungo Park was captured by unfriendly Arabs. After four horrid months he managed to escape, only to face a lone 3,000-km journey through the uncharted Dark Continent of Africa. He suffered hunger, fever, and was robbed of all but his trousers and hat before he lay down exhausted to face certain death.

"…at this moment, painful as my reflections were, the extraordinary beauty of a small moss in fructification irresistibly caught my eye. I mentioned this to shew from what trifling circumstances the mind will sometimes derive consolation… Can that being, (thought I), who planted, watered and brought to perfection in this obscure part of

*Left: In the heart of Southampton, a patch of gravel supports a mix of windblown or resilient colonists. **Buddleia**, **Pineapple Daisies** and **Mugworts** all combine to produce a colourful and natural city garden.*

the world, a thing which appears of so small importance, look with unconcern upon the situation and sufferings of creatures formed after his own image? Surely not! Reflections like these would not allow me to despair. I started up and disregarding both hunger and fatigue travelled onwards assured that relief was at hand."

It was. He arrived in London 19 months later only to return to Africa in 1805. This time, however, he was short of moss and he was killed by natives in rapids on the River Niger. What a way to go!

Apart from the romantic Mr Park, early botanists failed to notice the mosses growing almost invisibly under their feet. The famous Greek Theophrastus (300 BC) and the equally celebrated Roman physician Dioscorides (77 BC) both ignored them at the expense of most other plant groups and it was not until the fourteenth-century 'Herbals' appeared that a few types of Musci were first noted. By 1782, Johann Hedwig had produced some beautiful drawings showing the dissected sexual apparatus of common mosses, and put forward theories as to their function. In 1851, Wilhelm Hofmiester finally solved the sexual riddle of mosses by explaining the complicated alternation of generations. Apparently his detailed studies were aided by his exaggerated short-sightedness which permitted him to prepare the tiniest fragments of moss for microscopic examination.

Today, some 15,000 species have been described worldwide, exhibiting typical extremes of variation. The largest, *Dawsonia* sp., may grow up to 70cm high in wet, shady forests, whilst the smallest, *Ephemerum* sp., stretches no more than ½mm high. Virtually unchanged for 320 million years, mosses are simple in construction, lacking the specialized differentiation found in higher plants and often consist of no more than an organized central strand wrapped in a tube of cellular chaos and a sheath of simple leaves. There are no roots, just tufts of rhizomes which bind them to their substrate where they are either tuft- or carpet-forming.

Although our modern DIY supercentres stock little in the way of moss these days, in the past

we have found a great variety of uses for the stuff. In Roman times, floors were carpeted with *Hylocomium* sp., while the long, tough and pliable strands of *Polytrichum* sp. were plaited into caps, baskets or ropes which tied Bronze Age canoes together. Other mosses were used as stuffing for mattresses and pillows, for making dusters and lampwicks and for the decoration of fashionable ladies' hats.

However, it was the pillowy Sphagnum Moss that was most extensively exploited. The Chinese used it for treating eye diseases; Inuit tribes used it to pad their boots instead of socks and in the West it provided a highly absorbent dressing for wounds. Whereas cotton pads only absorb four to six times their own dry weight of water or blood, bundles of dried Sphagnum suck up to 16 to 20 times the amount. These absorptive powers are due to the many large, dead and colourless cells which fill the leaves. By the end of the First World War factories in Germany, Britain, the USA and Canada had supplied over two million

*Above: Aside from being the 'Butterfly Bush' of love, **Buddleia** has a great capacity to germinate and grow in the meanest of crannies imaginable. It is a native of dry, rocky stream beds and seems to be able to grow in almost any soil or nutrientless environment. Stare skyward beneath old brick buildings and you are likely to spot a sprig of strangled green. It will be Buddleia.*

Sphagnum dressings a month to the war effort. Each sterilized wad was enclosed in a bag of fine muslin and once applied could be left on a wound for two to three days, far longer than any conventional bandage. They were also cool and soothing. It was only the abundance of cotton and its clean antiseptic appeal that ousted moss from its valuable medical role in the Second World War.

LOOK... LICHENS!

Take a walk through an old city cemetery and study the patterns on the stones. Wonderful and abstract mosaics, jumbles of spots and speckles, occasional blots of bright colour all daubed by nature over nostalgic names and ages. These vandals mix their graffiti slowly but their artwork is guarded in these enclaves, the sacred stones sure to persist longer here than on buildings beyond the hallowed ground.

Lichens are a stable mixture of algal plant cells and fungal mycelia and the resultant body, or thallus, appears consistent in form, enabling each combination to be considered a separate species. The advantages to the fungus are certain, it receives essential sugars from the algal cells to grow and reproduce. What the alga receives in return is dubious. Perhaps it finds protection from extremes of temperature, light or moisture, or even some mineral salts leached from the substrate by the fungus. However, this is more than likely an example of parasitism where the fungus determines the shape of the thallus to maximize the alga's photosynthetic ability. An indication that this is probably the case is that most of the algal species found in lichens can and do appear free-living elsewhere, whereas the fungal partners never exist in isolation. The fungi cannot break down living or dead organic material and are entirely reliant on those plant-produced sugars from the algae.

*Above: Perfectly happy on rocks and hard places, lichens such as this **Physcia caesia** are nature's answer to the nastiest spots on earth. If it's truly extreme then only this mix of algae and fungi will dare to grow and prosper.*

Right: Xanthoria sp. Shrivelled lichens from ancient herbarium collections have been re-hydrated and nursed back to life with tap water. They can be all but indestructible so that even time cannot destroy them!

The mix also imposes constraints upon reproduction because only the fungi produce fruiting bodies and there is no provision for any algal involvement. Thus, after the spores have been dispersed and germinated, they have an essential need for plant tissue and some parasitize other lichens to steal their algal cells. Despite the massive numbers of spores produced, probability weighs against successful reproduction and consequently many lichens rely on vegetative means of dispersal and colonization. Special propagules develop on the thallus and these break off to be carried away by wind and rain or by insects and birds. If they fall into the right micro-habitat they produce a new lichen body. One species that presumably employs such means is *Xantheria parietina*, which often grows on bird perching sites. Its crinkley, brimstone yellow thallus can be found on rocks, walls, fences and old asbestos roofs in most urban areas because it is tolerant of air pollution.

Lichens do not have common or English names and are known only by their scientific names, possibly because identifying them to species level is a nightmare. Their colours change according to whether they are wet or dry. Their

microscopic spores are often a decisive but invisible characteristic and many species lack any real form, appearing as crumbly scabs or stains on their substrate. Only a few fanatics with microscopes and chemistry sets, who cut minute sections of the thallus and squash it onto a slide for examination, are able accurately to name our lichens. Standing at the bus stop, regarding a small grey encrustation on a wall, you might be forgiven for asking, what is the point?

STINGING IN THE RAIN

The next unloved city survivor is that provider of itchy welts, the ubiquitous Stinging Nettle. The ferocity of this plant is not only directed at the hands of gardeners, hapless children, blackberry pickers and a host of herbivores, though, because this is one of the most aggressive and competitive plants in Britain, inflicting cruelty upon other greenery through some extraordinary adaptations.

Put your gloves on and gingerly peer into the verdant canopy of a nettle bed during the height of summer. You will find the soil below the tall stems is quite bare because no other plants can stand up to this ruderal barbarian. It is a light hog, a species that outcompetes everything for this valuable resource and grabs all the nutrients to produce masses of leaf material. It is tempting to think that plants are of a fixed design, that once grown they remain the same, each structure performing its function in an unchanged state before it turns brown and dies. But, in the case of Stinging Nettles, this is far from true as I discovered when I investigated the structure of the plant and found that throughout its life it continually readjusts its leafy canopy to maximize efficiency. When I was a student, I grew a large bed of nettles in one of the university greenhouses (an unpopular move with staff and other students alike), regularly photographed my crop from all

Left: *Hurrah! Here it is: the marvellous result of our urban neglect, an example of a once rich series of valuable oases, a poignant, colourful, vibrant and beautiful patch of wasteland with* **Field Bindweed** *and* **Rosebay Willowherb**. *What's that song? 'You don't know what you've got till it's gone.' So we city folk need to protect our 'Brownfield sites'.*

angles and measured light levels and all sorts of planty parameters such as stem height, leaf area and leaf length and breadth. By the end of the summer I had a wealth of data, had cursed countless stings and had made a few discoveries.

Nettles are limited in terms of their leaf canopy design by the fact that all the leaves arise from one central stem, so they are stacked pretty much on top of each other. By the time the seedlings reach 15cm tall, there is an 80% overlap which persists until death. Only the top four or five leaf storeys receive any direct light and because these allow only 10% of the light they receive to pass through them, those below are obviously not concerned with the interception of direct top light. Thus, the plant has a problem. To visualize it and its solution go and sit in a bed of Stinging Nettles. Carefully!

As you look down from the top of the nettle you will see that the leaf stems, 'petioles', at the top are nearly erect, those in the middle near horizontal and those below actually allow the leaves to slightly droop. Most surprisingly, the shape of each individual leaf changes from the time it appears at the top to when it works its way down the stem as the plant grows. Leaves begin relatively long and narrow, immediately broaden and then become thinner again when they are several storeys down. As it turns out, all these developments mean that the leaves are

Left: *Not the property of a keen cyclist, these spokes are interspersed with the florescences of* **Common Ragwort**. *This ruderal colonizes and grows quickly on the poorest of soils, as do a host of other brightly flowered plants. It is a boom time specialist whose aim is to get in, seed and get out again.*

*Above: The sheer audacity! Here sprouts a couple of **Willow Trees** in defiance and disregard of all those ordered bricks. Thank goodness they're out of ladder range.*

appear at densities of over one hundred stems per square metre and finally grow over a metre tall in dense overlapping stands. Their dynamic system of adaptation is so good that in the end only nettles can realistically compete with nettles and all other plants shrivel in their shade.

THE PHOENIX PLANTS

Nettles are not the only ruderal refugees to invade our cities. Those linear gardens which edge our roads and railways, and stretch to stations and car parks deep inside the manscape provide a route for many plant colonists. Birch, Ivy, Bluebell, Primrose, Ramsons, Bramble, Gorse, Broom and countless other herbs and grasses carpet the cuttings, embankments, yards and verges. That opportunist commuter the Oxford Ragwort is reputed to have spread around Britain courtesy of the railways after its seeds blew over the wall of the Botanical Gardens at Oxford onto the embankment beyond. These plants normally produce between 10,000 and 160,000 tiny, windborne, parachuted seeds each year. It is generally accepted that the construction of the Great Western Railway in 1838 allowed the plant to spread on the breeze created as the trains swept along their tracks. The ash and cinder bases of the tracks also provided a more than adequate substitute for volcanic screes where this species occurs naturally.

Untended city plots tend to be barren, stoney or have little or no soil of any quality, conditions that have actually favoured other immigrants. Canadian Fleabane is thought to have arrived from North America in the Eighteenth Century and initially prospered on walls and odd patches of wasteland but it was not until the Blitz that this and another opportunist, the Rosebay Willowherb, exploded onto our urban scene. The bombing peaked between September 1940 and May 1941 and despite the tenacity of the population it was impossible to rebuild our city centres until the war had ended. Almost immediately, amongst the ripped walls and heaps of rubble, there was a rapid and colourful flowering of weeds, the most common of which was the Rosebay Willowherb. From pollen analysis of ancient soils we know that the species had always been a British native growing on open ground, eroded riverbanks or cliffsides but such habitats were never widespread so it was never common. But as soon as motorists began racing about the countryside casting lighted cigarettes to the verges the numerous ensuing fires provided

able to use both direct and diffuse light in quite a complex compromise.

The first fat leaves of the shoot expand to immediately obtain as much light as possible and get the seedling going while the canopy is still shallow. Later the upper erect and narrower leaves allow more light down into the deeper canopy where the then horizontal and broader middle leaves can maximize the use of diffuse light. The spacing of these and their angle continually changes to remain the most efficient trap for side-light and because these leaves are held away from the stem pairs of extra little leaves shoot out around it to collect any light leaking through these spaces. Such a vigorous effort to capture light is not surprising as nettles a superb new habitat for Rosebay Willowherb and by the time war broke out the plant was already creeping citywards. Each fruiting plant produces up to 8,000 fluffy seeds and these woolly clouds soon wafted in to colonize burned and barren spots in the cities. Subsequent examination of bombsites by the London Natural History Society revealed that most of the plant species colonizing them were those with windborne seeds and that there was a greater abundance of new plants on the west side of London closest to the prevailing winds. In 1939, only twenty or so weeds had been identified in the centre of the capital but by the 1950s more than 275 new species had sprouted in the square mile that surrounds St Paul's Cathedral. I suspect that fewer survive today but if you stand anywhere in the city and look hard enough you will be able to spot a sprig of feral greenery. It may be high up, struggling or hiding in a crevice but I can't help wondering just how quickly our urban jungles would be more jungle than urban if we all upped and left? We wouldn't have to wait long, I'm sure!

Above left: Silver Birch *is the pioneer tree – poor soil is no problem and it also caters for a great range of herbivores that munch on these tasty leaves. The buds and catkins also provide a great food source for birds, particularly the tit and finch families.*

Above right: *Split, crusty, gnarling, peeling and nutritious: the perfect substrate to colonise. Lichens, mosses and fungi all have fun at the expense of* **Silver Birch** *bark and its crannies provide a home for hibernating or hiding insects.*

Right: *Iron will become rust and then dust in the end. Here the peeling paint of the railway bridge provides a poignant backdrop for the bouquet of verge-side vegetation, which stands as a totem to nature's perfection and permanence.*

CITY SAFARI: MAMMALS, REPTILES, AMPHIBIANS AND FISH

We can't seem to moderate our obsession with celebrity and the exotics. The following panders to this fixation for the beautiful, the popular, peculiar, the friendly and familiar. Of course, I can't resist including a few of the 'C List' underlings. So let's say a big 'Hello', give them a bit more exposure, a modicum of hype and the right friends, and they might just make the T-shirt.

THE TIGER BEATER

A quick shadow slips behind the tidy bin and after a few seconds a pointed nose emerges from the other side. The muzzle is followed by the brightest of eyes and keenest of ears, then a shaggy neck and long straight legs with delicate feet that shuffle on the tarmac. When the Red Fox turns to face you its true beauty is revealed. Contrary to widespread mis-information, city foxes are not thin, tatty or mangy versions of their healthy, bushy, country cousins. I've seen some of the most pristine, fully furred and attractive members of this family in my headlights or trotting down my driveway. But then it appears that the urban fox is enduring the same fate as so many heroes – we build them up to celebrate them and then knock them down and destroy them with equal enthusiasm. You see, 30 years ago we were all amazed by and enamoured of the widespread success that these creatures had cunningly achieved in our towns and cities. *Blue Peter* ran stories about the phenomenon. To great acclaim the BBC produced a nosey and nightly programme, *Fox Watch*, which detailed the exploits of a family of Bristol foxes, and Brian Vesey Fitzgerald penned *Town Fox, Country Fox*, a popular book which investigated the behavioural, physiological and ecological differences between the two populations.

In truth their increase has continued and Britain has more urban foxes than anywhere else in Europe. Within the M25 there is thought to be a population of some 5,000 animals and some of the leafier boroughs of the capital with older, bigger and quieter gardens have very healthy fox populations indeed. The key to their success, as ever, is food. In cities we leave lots of it lying about in litter bins, dustbins, put out for birds or spread on the compost heap and of

Left: Red Fox. *Those were the days – galvernized dustbins with their light, easy-to-tip and rattly plastic tops. Now the safe, secure and silent wheelie bins have ruined everything. Anyone for a classic bin revival club?*

Right: King of the shed. What a beauty, statuesque and superb, caught in a moment of study. This photograph provides an apt antidote to the idea that the **urban fox** *is a tatty, mangy, pale shadow of its rural counterpart. And it's lucky enough not to be hounded across the countryside.*

course deliberately laid out by people who enjoy the company of these adaptable canids. And they'll eat anything – pizza, curries, Chinese, Thai – if you've taken it away they'll finish it off, even the dreaded donner kebab will have a fox licking its lips and it won't need to be drunk either! But even without our slovenly littering they do well on a diet of other urban wildlife from worms and fruit, to mice and Starlings, their predatory skills have not gone soft with city life.

But while the foxes have continued to prosper, their popularity has been on the wane, fuelled by inaccuracies on all sides. In suburbia they have been repeatedly and wrongly accused of killing cats, small dogs and other family pets. And yet, there are no substantiated accounts to back up this charge. Indeed, most observers report that foxes are subservient to cats and clearly fear dogs, even those perverted pedigree pooches that are more like rodents than canines! 'Disease-spreading pests' is another vitriolic fallacy. Rabies

Right: Carpet underlay and a crumbling shed are paradise for this pretty **vixen** *who was, until stirred, slumbering in the sunshine. This little quietude is all a fox needs to get away from dogs and doze away the memory of rearing cubs.*

Left: *The **Common Shrew** likes to spend its life in the security of the deepest grass, rarely revealing itself to the wider world where it would fall prey to raptors et al. Thus, we rarely see it, but as it's fiercely territorial and squeaks maniacally during disputes, it can often be heard by a trained ear.*

has yet to become established in the UK and no longer warrants the fear that it once instilled due to advances in vaccinations and treatments for both animal and human victims. Mange has become widespread in city foxes, an artefact of their relatively dense and stable populations, but there is only a very slight chance that it might be passed onto your pet because, as I've already stated, foxes eschew contact with them at any cost. It seems sad to have to point it out, but this horrible disease cannot be passed to humans, a theory that I have heard voiced. In reality, we have far more to fear from our pets: cats and dogs greatly outnumber foxes in the UK, and there are those devils among us who still refuse to 'scoop the poop' and therefore expose us all, and particularly our children, to the *Toxicara* parasite. This worm still causes blindness and illness in the UK with more regularity than any other fox-transmitted pathogen. Ask yourself, when was the last time you slipped on, trod in, or turned over a pile of fox poo on the sports

Left: *The **Bank Vole** is principally a woodland or hedgerow animal (not an open grassland species like its relative the Field Vole) and is suited to life on railway verges, park borders and in large gardens. It is active day and night but prefers the latter as it is easier to avoid predators. Keep your cats indoors and give the vole a chance.*

*Right: A **Pipistrelle Bat** on the attack. In the summer dusk, insects are the targets, anything from midges to moths. This high-speed picture shows us what this remarkable creature looks like when airborne. Soft membranes stretched between fragile fingers, fixed to a hyper-mouse body, do the trick. We see flickering silhouettes; the camera reveals the science.*

*Right: A cosy community of **Long-eared Bats**. They are one of the few species that can really adapt to urban life. Most of the others need to feed over water, but these nocturnals pick insects from tree foliage in a similar fashion to flycatchers. They'll eat anything from bugs to beetles.*

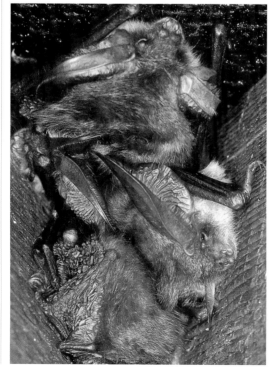

field, pavement or garden border? Thanks to the wheelie bin, any nuisance caused by late night littering is largely a thing of the past and if a bit of nocturnal wailing for a few weeks of the year annoys you, wouldn't the purchase of ear plugs be a little more appropriate than pestering the council about pest control? You see, for foxes, finding a partner, courting and mating is a noisy affair, but never at quite the decibels you tolerated at the disco or nightclub with similar objectives in mind!

Dog foxes maintain territories that are larger than those of the females and typically overlap with two to four of the latter. Their problem is that when those guarded females come into season, a short-lived affair, rival males from adjacent territories will attempt to sneak in and score. Consequently, fights break out and the characteristic snarling, yelping, whinnying and screaming all reach a climax. Which is fine, if it isn't taking place outside your window all night! Still, because the males need to rove about between the females' ranges, such impromptu concerts are rarely repeated at the same venue.

Research has shown that urban foxes live an average of just 15 months, although some senior citizens may make six or seven years. It might be that cities offer rich pickings in terms of food but there are plenty of perils in the package, not least traffic. Despite their nocturnal habits which see them active when there are fewer vehicles on the road, these days the streets are rarely quiet for long and as your own kerbside censuses will confirm, there are lots of casualties.

I'm a great fan of the urban fox – for me a winter-coated animal, wrapped in a thick glossy red jacket, with a fluffy brush, black socks and so delicately drawn with white, is every bit as beautiful as any tiger. And it is not in a distant jungle, flickering on television or staring from a poster. It is sharing my space and I'm not going to be fooled by a load of urban legends that seek to undermine the success of this streetwise survivor.

CHAMPION ON YOUR NUTS

For all its glory the fox is not the most frequently met mammal in the metropolis. That honour falls to the greatest invader the UK fauna has encountered, the one and only, the definitely indefatigable Grey Squirrel. Introduced a little over 130 years ago, it has a very firm grasp on its place on the British list and has spread everywhere short of northern England and Scotland. Sadly, it is fairly universally despised, blamed for the demise of our 'own' Red Squirrel, targeted as a pest in tree plantations and even loathed because it steals the eggs and young of 'our' woodland birds. But such perspectives are infused with inaccuracies and not a little peculiar racism. The Red Squirrel is not so strong a species, has been blighted by its susceptibility to disease, happened to be at a low when the Greys arrived and so lost out to a better adapted competitor. There is no doubt that the Grey's bark stripping habits and taste for new shoots mean that in places where it is common it can be a nuisance to arboriculturalists but how this reputation has been carried to the park or garden is a mystery. Or is it more eco-racism? And as for pilfering nestlings, well yes, of course 'Squidgy' is guilty from time to time, but not nearly so effective as our 'own' Pine Marten would have been if we had not all but exterminated it. Look, the Grey Squirrel's greatest crime is the chewing of our nuts, those that we put out for 'our' birds. I find them cheeky, friendly, entertaining and very successful; they should be enjoyed and championed.

Park Bandits

There are countless city centre parks where Grey Squirrels have become exceptionally tame and delight young and old alike. In Bournemouth a quaint path winds from the town centre to the pier through manicured lawns and beds. It is lined

Left: *It might be our most obvious invader, but, let yourself succumb to the charm, and check out the **Grey Squirrel**. Forget your patriotism – the Reds have gone for good; and all those stories of Greys nestling, eating and tree crippling are exaggerated. Besides, there are bigger issues in wildlife conservation than the non-native naughtiness of the Grey Squirrel.*

Left: *The **Grey Squirrel's** tail is used as a means of quickly communicating with other squirrels. Flicked, shivered and flattened, the tail conveys messages.*

with benches permanently supporting the weary limbs of the aged with whom the place is synonymous and who are all entertained by some very clever cads indeed. No sandwich is safe, no lunchbox sacred! But recently the council outlawed the feeding of squirrels. 'Do not feed the squirrels' is how they put it. Apparently, they can bite, and this could be dangerous. They'd do better to worry about some of the genuinely life-threatening events that take place outside the city's many nightclubs than fret over a little finger-nipping-naughtiness in the park. Holland Park in London also has a charming contingent of squirrels seemingly bent on purloining the top-notch picnics available in this 'nanny-land'. So adaptable is 'Squidgy' that he can turn his taste from nuts and shoots, from corned beef and tomato sauce to *fois gras* pâté, chilli-olive *tapanade* and Fortnum's baby food! Take my advice – wherever you are, turn your scowl into a smile because the Grey Squirrel will always be one of life's winners.

BLACK AND WHITE BLIGHTERS

Far more popular but far more discreet is the urban Badger. Entrenched in dark rural secrecy, this animal has become an icon even while it remains invisible to most of its fans. Its cuddly, mumble-bumble and amiable lifestyle has made it top of the mammal pops. I'm afraid, however, that its reputation is all spin and the Badger is actually a bit of a blighter.

Badgers are cautious animals, shy and relatively secretive, but they have adapted very well to life on the fringes of our towns and cities through one essential aspect of their ecology. Their choice of food is even wider than that of foxes. They are omnivorous, eating vegetable matter as well as meat. They will eat acorns, pignuts, crab apples, beechmast and fungi, in fact, almost anything if it is green and grows. They also prey on rabbits, birds, beetles, wasps and earthworms – and this is the key to their success. For all their cosmopolitan tastes, Badgers are earthworm-specialist feeders. Nothing matters more to Badgers than worms

Left: Rabbit. *Thumper, Peter and all the 'rabbity' rest amount to a lot of prey for a lot of species, which are far more interesting. Bunnies are food-chain fodder.*

and where to get them, and our own taste for lawns and sport put worms on a plate for them. The generous scattering of closely mown and tended garden lawns, park greens and sports pitches, make suburbia a better place for Badgers than some of these creatures' rural retreats, particularly extensive conifer plantations, heathland or downland. There, worms are rare, but sniffing over the studmarks between the goalposts, Badgers have no trouble finding the oodles of slippery noodles they require. Their nose goes down, flexes up and their little lips purse on the soil to suck up the worms. I've heard them eating – very poor manners – wouldn't be popular in Holland Park!

The whole of the Badger's ecology is governed by this dependence on a single food source, which influences their distribution, size of territory, the number in their social group and their fecundity. In truth, urban Badgers do deviate a little from this model but still show a keen focus upon the availability of worms and without grasslands no amount of dietary diversions will tempt them into town. Of course, apart from food, they do require

homes to go to and here peace and quiet are prerequisites. Thus, wasteland, untended parkland or large 'un-gardens' are sought for sett building. These tunnels, chambers and holes are never as extensive as their countryside counterparts but because freedom from human or 'doggy' disturbance is essential, finding a plot is difficult. One habitat that is invariably available for refuge is that provided by railway cuttings but this has a definite down side for house-hunting Badgers. Their legs are not long enough to clear their bellies above the electrified third rail and each year hundreds of Badgers perish trying to cross railway lines. Nevertheless, pioneers do persist and in many places this species has a secure toehold in urbania, enjoying a generous top-up from increasing legions of late night Badger feeders. The sight of several of these creatures shuffling across the lawn sure beats *Newsnight* or Snooker. A clear hierarchy will be evident amongst the diners, something which gently betrays the fact that although these animals may be charming, they are not cuddly. Strongly toothed and clawed, they are fearsome

*Above: Like so many animals, the **Badger** prefers routine, so if you feed it regularly with a sweet and tasty menu – jam sandwiches, biscuits and Sugar Puffs – it will come back again and again. Sometimes, however, only the slightest change in your pattern of practice is enough to spook it, as even in towns, it retains its wild nervousness.*

*Right: The **Hedgehog** is not as common in urban or suburban areas as it used to be. Too much traffic and too many slug pellets are contributory problems. In the thick, grassy and brambly patches of rail verges, it still does well, but if you've got one at home, look after it.*

adversaries when they indulge in disputes over territories or drive young animals from their home setts. A friend of mine has a sett in his garden under constant observation and a few years ago the residents quite literally and inexplicably fought each other to death, to the point that the remaining Badger left, and what months before had seemed a stable community had suddenly committed ritual suicide. So when you next pull on a Badger-fronted T-shirt, dust your porcelain Badger ornament, or pin up a poster portraying this wildlife idol, remember – behind the furry façade lurks a worm-slurping psychopath – nice!

*Below: The sensitive and secretive nature of the **Roe Deer** means that most people are astonished when they see one in town. In fact, they are quite confident commuters so long as they have a little patch of impenetrable quiet and are able to go unseen due to their cautious and nocturnal nature.*

*Left: The **Great Diving Beetle**, the bruising thug of the pond. Its larvae are fierce beyond compare and the highly mobile, winged adults bring death to any dirty wasteland puddle they arrive in. Like so many species, they roam and thus are capable of finding their way deep into our cities.*

WHERE THERE'S MUCK THERE'S LIFE

It's sad that people no longer dump prams into canals, bicycles into ponds or drive Ford Anglias into those greasy slicks behind industrial estates. This rusting litter displayed the great art of decay, demonstrated the impermanence of our technology in contrast to nature's endurance and hid huge populations of those creatures that would lure bare-footed young naturalists from the bank. A washing machine once yielded a Great Diving Beetle for me and numerous newts poured from a de-galvanized dustbin. What appeared from beneath a car seat is best not recounted! Sixties litter was great; now the plastic prams go from boot sale to boot sale and only the occasional supermarket trolley reflects shimmering patterns onto the stagnant green of our urban backwaters. Gone is the glory of naughtiness, of entering tetanus-filled lagoons, the forbidden dirty places where adults feared to tread, to recover decrepit hulks of bikes. These frames scratched pavements all the way home while you fantasized of restoration only to see your mum reel in horror and wince at the bitter tang of the mud oozing from the jammed handle bars onto the kitchen lino. Dad took them to the council dump cursing your folly and thus dreams were lost to landfill. Now the kids have got computers, won't get dirty, cut, beaten up by bigger boys or tempted to dream of something from nothing with a tadpole and newt as a bonus. I meet these kids and they're dull. They're as dull as the ditch-water they never fall in.

However, it is not just the sculpture that has been stripped from our urban streams; most of the pollutants have too and the water companies and the Environment Agency continue to mop up the mess and penalize those who still see such waters as a sink where all evils will dissolve away. Fish, real fish, with silvery sides swim in the trickles that twist and turn between our buildings and bigger fish cruise in the canals and rivers, which run through the hearts of our older cities, no longer choking on chemically corrupted currents. It is not quite a crystal clear pristine paradise, it never will be, but there's no doubt that we've cleaned up our act. Still, complacency is not yet the order of the day and we need to maintain a strict focus on the health of our waterways particularly as this essential, basic stuff of life seems to be draining away from our countryside. The future is not orange, it is wet for wildlife and for us too!

THINK AT THE SINK

Generally, we have a medieval view of drains: they are for washing away all the world's waste and we pour all sorts of nasty things into them: paint, oil, detergent. You name it. If it's nasty, it goes down the drain. But some urban drains feed little oases of marsh, mud and murk and they don't need topping up with toxins. So think twice when you're standing at the sink, especially at work, where worse things than cold tea are poured away.

TOADS VERSUS TYRES

Such improvements are welcome but must be countered with the ongoing loss of many waterways and ponds, an effect particularly prevalent in urban environments. In a 1985 London survey 90% of water bodies present 120 years before had vanished, a fact which has severe consequences for those species that require connected, permanent water sources. Colonization or re-colonization following local extinction is becoming difficult or impossible. Indeed, even the yearly migration back to their breeding waters is increasingly difficult for our amphibian species. One relatively recent social development has been the instigation of yearly 'Toad Patrols'. Monitored and measured by the charity, Froglife, these comprise groups of volunteers who collect toads as they approach the road and, avoiding the ever-present traffic, release them safely on the other side. It sounds simple but predicting toads' movements requires some practice and local knowledge and it can be a gruelling, cold, wet and time-consuming task. Spread across Britain there are about 400 spots where toads jay-walk in large numbers and the most perilous of these are on those routes in and out of cities where traffic is heaviest. The slightly eccentric nature of the endeavour appeals to the local press on an annual basis, and, of course, safety is of the utmost consideration, particularly if bedtime-avoiding children are in action. Froglife dispense reflective tabards and roadsigns, so if you identify a slippery stretch of toad-strewn highway, ring 01986 873733 for advice sheet No. 3 and try to lure some of your friends away from the hypnotic tedium of early evening light entertainment.

Still, away from the road there is some good news for those amphibians that can prosper in garden ponds because even the smallest of tubs can provide enough secure water for their success. Frogs are the primary users and find a vital refuge in the typically shallow ponds that are surprisingly numerous in suburbia – Greater London has as many as 150,000 ponds. They also provide a home for Smooth Newts and Common Toads. Tellingly, all three species are now doing better in urban areas than they are in the countryside and the population of frogs is probably still increasing in our towns and cities. Toads do not fare so well because they prefer deeper water in which to lace their spawn strings, something you should consider when designing your next water feature. Indeed, even the Great Crested Newt, a threatened species that is therefore given special protection, can be tempted by deeper water. I once helped evacuate an ancient swimming pool in an overgrown garden and it was alive with these mini-dragons. They were all safely relocated but the 'pond' was filled to provide a car park for the low-rise block of flats that replaced the wonderful old Victorian villa. And it is also this 'infilling', so grossly popular with developers at the moment, that most seriously threatens our few remaining sites for urban reptiles.

Below: Common Toad Tadpoles. An amphibian metamorphic stage, maybe, but there is no denying that the tadpole was designed with the jam jar in mind. Cute, cuddly, easy to catch, easy to feed, keen to wriggle and nibble, if only we could harness the happiness these creatures give to legions of children every year!

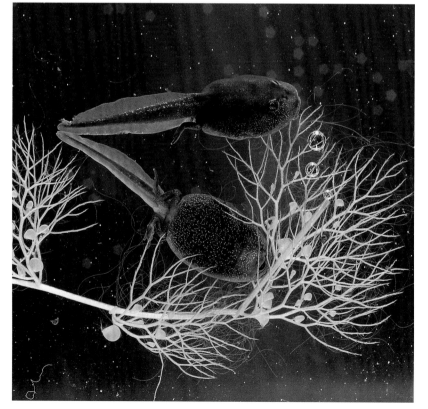

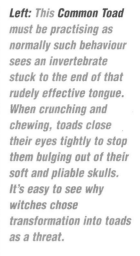

*Left: This **Common Toad** must be practising as normally such behaviour sees an invertebrate stuck to the end of that rudely effective tongue. When crunching and chewing, toads close their eyes tightly to stop them bulging out of their soft and pliable skulls. It's easy to see why witches chose transformation into toads as a threat.*

*Left: **Smooth** or **Common Newt**. Despite being another jam jar celebrity, the newt is not renown for its dynamic and interesting personality. At least males, like this one, become spotty and crested in the spring and show some interest in the females. Aside from that, newts look good but do nothing, an adequate parallel with many human celebrities.*

*Above: One of nature's track and field champions in action. This **Common Frog** demonstrates the reason for having those long legs, displaying behaviour that might save it from being 'catted' just so long as it lands in the pond. Imagine the fun you could have if it was trainable – flaming hoops, motor cycles, and other daring displays!*

SCALY SLICKERS

When I was a child 'wasteland' was widespread throughout the city of Southampton. A corner here, a strip there, a block behind seemingly everywhere; and each patch was an oasis, a refuge, a home for so many species. Today it has all gone. Tiny houses or ugly flats have systematically filled all of those patches, the driest of which were a superb habitat for Common Lizards and Slow Worms. The delight of flipping over an old fridge top, or, after a helping heave from a mate, a length of corrugated iron to pounce upon a sleeping Slow Worm has gone from the lives of our cities' kids.

Common Lizards and Slow Worms both have diets and lifestyles which enable urban living. All they require is a little cat-free space and some quiet. English Nature estimates a UK population of 1.2 million Common Lizards – notably declining in the south and east – and 660,000 Slow Worms – decreasing rapidly in Scotland and the East Midlands. Coincidentally, both the Adder and Grass Snake are also declining in the UK and since the 1980s have disappeared from many of their urban haunts. Allotments and larger mature gardens are the best hope for all of these 'scaly slickers' but these plots increasingly catch the greedy eyes of developers and sadly, current legislation means that local councils are powerless to prevent or even slow the insidious creep of 'infilling'. What a sad price we pay to park our cars and live on top of one another.

*Left: When **Common Lizards** are active on the dryer 'wasteland' in the city and suburbs, between April and September, they can be remarkably trusting. They are naturally curious, and, if carefully approached, will allow food to be offered – flies and grasshoppers are a favourite. Even hand feeding them is a possibility with some individuals.*

Up the wall

There is one species of reptile that seems to be doing all right though, but it is probably a non-native and therefore its very presence is frowned upon by those eugenical purists in conservation. The Common Wall Lizard has established itself, or perhaps more pertinently been established, at about twenty places as far north as Shropshire, including several sites within cities. Its stronghold, at Ventnor on the Isle of Wight, is less than metropolitan but provides a perfect example of its continental requirements. It is not afraid of the cold, occurring naturally in Holland, Belgium and northern France, but it does like its shelter and sunshine, something it finds on

*Left: If mustard is no good without roast beef, then rusty corrugated iron is no good without a **Slow Worm** or two. This declining, slug-eating reptile capitalizes on the thermal qualities of this material nesting under it for safety and warmth.*

Above: Grass Snake. Only a few of these reptiles survive in our towns and cities as they need water, quiet and an absence of cats and dogs. In the 1970s, a colony prospered in Southampton city centre but they seem to have disappeared due to disturbance.

Right: Once warmed-up, this alert and super quick little Wall Lizard is almost impossible to catch. Before it reaches operating temperature, it is vulnerable to predators. The Wall Lizard does at least enjoy the sanctuary of the vertical and seems to be doing well in our run of mild winters.

our buildings. It likes to climb on walls, parapets and path sides, or anywhere dry where cracks, crevices and crannies provide bolt-holes, and throughout its range, it displays a fondness for human habitations. A friend of mine maintains a marvellous colony on his cottage. Where the south side catches the sun, these lovely variably brown-grey and long-tailed lizards scuttle skywards with astonishing alacrity whenever they are disturbed about their fly-catching business. At the moment, Wall Lizards show no signs of expanding their invasion, perhaps reluctant to leave their brick cliffsides and stray into the realm of the rotten old cat. But if our current run of kinder winters continues perhaps some may commute to our city centres where the extra few Centigrades would be a welcome bonus to them.

STOIC SWIMMERS

Even before we took such a determined stand on pollution, many urban waterways managed to support a fish fauna, albeit reduced and invariably including the Roach. This species is a favourite with anglers and, consequently, its presence or absence has always been known and because it tolerates moderately polluted water and low levels of dissolved oxygen, typical of stagnant canals and back waters, it has always swum in our cities. It is a member of the Carp family, which dominates British freshwaters and also includes Carp (obviously), Tench, Chub, Gudgeon, Bream and Minnow. The Roach is not a monster of the deep, rarely growing longer than 40cm or weighing more than 1kg, but it is a good-looking fish, having a dark, rounded back, silvery white sides, orangey eyes and characteristic rich red fins. Like all good urban adaptees, it has a varied diet. Young Roach feed on crustaceans and water fleas and as they grow consume more insects. Thus, mayflies, caddis flies, midges and their larvae, Water Boatman, water beetle

larvae and even pond snails and Freshwater Shrimp find their way onto the Roach's menu. It will even eat algae in the spring and summer and, as a consequence, there's always something to suit its palate in the stream. Roach

*Above: Even from the sunny surface of a river, a shoal of **Roach** is a well-hidden group. From the top, these otherwise splendid fish appear dun and dull, their counter-shading and camouflage blending in perfectly with the wrapping weeds and shifting silt. Only the keen-eyed Grey Heron will spot the shoal.*

*Left: Gammarus, the **Freshwater Shrimp**, one of the most extraordinary animals in all of God's universe. When I'm re-incarnated, I want to become a Gammarus living in a stagnant slack behind a tyre and exhaust centre in Walsall.*

Right: Anything that can breathe through its bottom deserves admiration, and, if it's a big-eyed, fast and vicious predator, then the points tot up. City ponds and puddles provide a hunting ground for this gladiatorial gobbler. The **Water Boatman** *can also fly and sing most Elvis Presley songs including the all time classic, 'Always on My Mind' – it's the ultimate all-rounder.*

Right: Caddis Fly larvae cover themselves in fragments of the substrate that they live upon. Here a few twiglets would successfully hide this one from predators such as Diving Beetles, dragonfly larvae and fish, if it hadn't been netted and put in a nice clean tank to photograph. Stones, sand, leaves, bark, twigs, even bits of plastic are delicately stuck into a safety tube by these resourceful little insects.

can spawn in colder water than many species and also produce a huge number of eggs. Thus from late March to May a large female will lay up to 50,000 eggs, giving the resultant fry a head start on those of other species with which they compete for food. But when it comes to curious breeding habits nothing beats the Bitterling.

This species began its British life as a popular component of cold water aquaria in the 1920s, but by the 1940s had established itself through multiple escapes and releases into the still waters of the Midlands. Being a common continental pond dweller, it fared well and can now be found in ponds, lakes, slacks, flashes and especially in the slow-moving canals and drains that lace through much of urban Britain. A giant measures 10cm from nose to tail, but in the breeding season, males are brilliantly coloured, transformed from a dull, silver-

Left: *The* **Bitterling** *is one of a huge number of non-natives that have found a niche in the manscape. Because the Bitterling is a fish and most of us never see it, this particular species fails to excite the hatred displayed by the more eugenical ecologists who stilal think there's a chance we can win back paradise. I'm of the opinion we should enjoy what we've got – wherever it comes from.*

brown fish into a green-backed, rainbow-sided, red-finned beauty. The females may be a duller version but become a more interesting animal in April or May when they develop a long coral pink egg-laying tube, which projects from the front of the anal fin. You see, the Bitterling's breeding behaviour is inexplicably linked with the Swan Mussel and the female uses this tube to lay her eggs inside the mollusc on to its gills. Cleverly, the fish accustoms the mussel to her presence by nudging it with her snout so that when she begins laying her two or three eggs the mussel doesn't snap shut on her delicate tube. After the eggs are laid, the attendant male Bitterling swims by and releases some sperm, which is sucked into the mussel's gills as it respires and thus the eggs are fertilized. They hatch after about three weeks, spend a couple of days inside the mussel and then make their way into the wider waters outside.

It's a superb natural plan. The mussel's breathing current is oxygen-rich, a bonus in any still water situation. If there is a drought, the mussel moves on to a more favourable spot carrying the eggs, and clearly they are well protected from predators inside the shell of the mollusc. And the mussel benefits too, because when the adult fish are busy breeding, it releases its own larvae which attach to the fish via a long sticky thread and eventually encyst into their skin. They are effectively parasites but do little to harm

the Bitterling, which in turn disperses the Swan Mussel wherever it swims. Brilliant, and it is all going on in that muddy ditch behind your office!

Left: *Try to imagine being a* **Swan Mussel**. *Drink eight pints of strong lager, sit in a bath of cold water and close your eyes, then drink a mug of mud and don't move for months. At the same time, think yourself lucky that no fish will lay eggs inside you. You see, for all the wonder of life in all its many forms, no matter who, where and how you are, I know you are better off than a bivalve mollusc.*

CENTURY 21 AND BEYOND

When I was a lad, 'Century 21' was a comic full of Thunderbirds. Now we've woken up to it and all the pressures it imposes on the environment. Unfortunately, the Tracy family are not going to help us, so I believe that the future lies in constructive and creative community conservation — that's us, me and you sorting our patches out. No puppets, just a few 'Saturday nights, Sunday mornings'.

A NEW BEGINNING

Throughout this book, I have enthused about species that have demonstrated an aptitude to adapt to the unnatural state of city life, often relentlessly praising their ingenuity or energy, always enjoying their success at whatever cost to our own. It is a case of championing the underdog – not very common in ecology. However, not all of our urban wildlife has needed to adapt or be damned: many species simply occupy habitats identical to those in rural areas except that they lie within the city boundary. Less manicured parks provide woodland and scrub for their typical complement of species, Hampstead Heath in London being a prime example. Householders with gardens put out food, erect nest boxes and the more ambitious actually plant with wildlife in mind, all significant benefits which enrich both the density and diversity of other species in our conurbations. Recently, there have been a number of projects on a grander scale. Two notable successes are the award-winning London Wetland Centre at Barn Elms and the work of the Sheffield Wildlife Trust.

A WILDLIFE TRUST SUCCESS

Great benefits can be forged for wildlife despite the apparent hostility of the environment, but it is also worthy to note that implementing such changes can also have extremely positive benefits on the human community.

Sheffield, sadly, is not renowned for its inner-city splendour. With the collapse of the steel, coal and manufacturing industries in the worst of Thatcher's 1980s purges, the city was blighted with problems. Crumbling estates such as Manor, Castle and Wybourn, once hailed as model programmes for modern living, became neighbourhoods rife with the worst of our social problems.

Rehabilitation is not complete but it has been aided by the inventiveness and energy of the Sheffield Wildlife Trust, who recognize that some aspects of our lives can be enhanced through the encouragement of nature. After surveys, audits and consultations, the Trust drew up a business plan which netted no less than two million pounds for a series of 'Green Estate' programmes. Overall, the scheme is ambitious but its components are

necessarily simple and already the results are profound. Areas cleared of tower block slums are now ablaze with wildflowers and ruderals, which profit from the disturbed ground. Species such as Cornflowers, Corn Marigolds, poppies and toadflax not only look good but provide a valuable refuge for all the usual suspects – bees, bugs and birds. These are not permanent features: new homes will be built here, but as the city slowly comes down to go up again these oases of colour will shift from site to site.

The manpower for these schemes has come from the community, so that those who sow are those who reap. Initial signs are good with a number of volunteers profiting professionally from a tree nursery business, which has been set up as a self-sustaining commercial exercise. Other community gardens are flowering on sites previously fly tipped and monopolized by intravenous drug users and glue-sniffers. The litter of abuse has gone, and ponds, orchards and a sensory garden host den-building kids and chatting pensioners. But perhaps more valuable still is the sense of pride and belonging that the local people feel as they slowly set Sheffield on the road to recovery through sweat and old fashioned toil. And, of course, behind the scenes the wildlife is smiling, too

Above: Wow! What a show! Imagine a misty dawn seen from the bedroom window of one of those semi-detached houses. Of course, poppy fields are short-lived affairs. A couple of years of splash and then the grasses get a grip and earthmoving is required to get the palette up to colour again – something the people of Sheffield have enjoyed to great effect.

THE LONDON WETLAND CENTRE

Below: A Bittern stands stock-still in the reeds waiting for a frog to flinch or a fish to flicker. Now that our capital has a habitat to support these national rarities through the winter, it provides real hope for the future of urban conservation.

Left: More often heard than seen in summer, Water Rails increasingly delight hide-hounds in winter by stalking out of the reeds to put on a show.

This 50-ha site was developed from scratch by The Wildfowl & Wetland Trust, Thames Water and, notably, Berkeley Homes on the site of a disused Victorian reservoir complex. At the project's outset some wholly unwarranted controversy arose because part of the site was given over to luxury housing. But this contributed in the region of £11 million towards the restoration of the remainder into a fabulous mosaic of wetland habitat types. Large lakes, pools, ponds, ditches, dykes, water meadows, marshes, reedbeds and willow woodland already provide a remarkable refuge for more than 180 species of bird.

Design work began in 1992 and a team of civil engineers, soil scientists and landscape architects sat down in a unique position – the 'book' before them was unwritten, no one had previously transformed over 40ha of concrete reservoir into a diversity of wetland habitats and a centre of educational excellence. After planning permission was granted in 1995, the awesome task began, primarily with the relocation and sculpting of over 500,000 cubic metres of soil, initially to seal the whole area and then to create the thirty plus lakes. Implicitly important was that each lake have an independent method of controling the water level, essential to maintain such fragile habitats, and to this end 27 sluices were built. Now given that no material

could be imported or exported from the site, the huge tonnage of concrete had to be re-used. It forms the base of over 3km of pathway, has reinforced shorelines, provides a refuge for fish in the deepest lake and underlies the car park. Naturally such sustainable use was an essential requirement in the plans and the 2,500 square metre visitor centre is made from oak grown in managed forests. Planting began in the winter of 1996. Twenty-seven thousand trees, and 150,000 plants and seeds, specially harvested from other Trust reserves, were bedded or sown on a site that looked very bleak and raw. At this stage, I made my first visit to the site and stood chilled in each of the seven hides that looked over barren mud and a motley-crew of confused Canada Geese and Carrion Crows. By May of 2000, reality had grown from all the fertile work and the centre was opened.

It is just 6.5km from the centre of London and already provides a home for ten pairs of breeding Lapwing, six pairs of Little Ringed Plover, 73 pairs of Reed Warbler and ten pairs of Little Grebe. Caspian Gull, Hen and Marsh Harrier, Honey Buzzard, Goshawk, Crane, Nightjar, Fulmar, Red-necked Grebe and Cattle Egret are all on its list and in January 2002, three Bitterns turned up. These were the first seen in London for more than a century and to have three in the capital is unknown in birding records. Other recent highlights include nationally important numbers of Gadwall and Shoveler, and 'cameo' appearances of Scaup, Red-breasted Merganser and Goldeneye. In autumn, the site plays host to huge flocks of migrating House Martin and Meadow Pipit as well as a marvellous population of Water Rail, which regularly show themselves to an increasing number of visitors.

Of course, not only birds have benefited from this habitat creation. The full gamut of aquatic life is flourishing, including a rapidly increasing number of dragonfly species, 19 recorded, 11 breeding, including Lesser Emperor and Red-eyed Damselfly. I attended a bat-walk here at the end of summer and the place was bat-tastic! Noctules appeared first, bending high above the network of paths and they were followed by a staggering number of Pipistrelles and Daubenton's which sent everyone's bat-detectors into a frenzy of chirps and clicks – all this set against a backdrop of the glistening city skyline. Fantastic! Admittedly, the weather conditions were perfect for such a display, but it was a fair rival to any you might encounter in a pristine wilderness and a fitting testament to the project's early success. Indeed, staff were so confident that the habitats were stable and secure that in May 2001 a project to introduce Water Voles was implemented. By October of that year live trapping revealed that the rodents were thriving and had even bred on the site.

A model, an inspiration and a challenge!

Now, if all this sounds like an unprecedented advertisement for the London Wetland Centre, it is. And why not? It is a triumph of creative conservation and has set a precedent as to what can be done in urban centres across the world. It is not just a resource for Londoners; I commute there from the South Coast, and in future it will draw visitors from all over the country. If you have not been – go, and then think about how you can replicate some aspect of its success where you live, on however modest a scale. Whether you work on your garden, a local park or 'wasteland', or some grander scheme, you will make a valuable difference.

If there is one message that this book hopes to carry, it is to admire the tenacity of life, particularly that which successfully overcomes our inconsiderate and selfish impositions on the environment. All that is needed by many species is an inch of space, an ounce of thought and a gram of tolerance. There is a philosophy in conservation that dictates that our responsibility should be motivated on behalf of our children or grandchildren so that they may enjoy what we do. But, to me, this embraces an arrogant conceit that the principal value of conservation is human gratification, a confirmation that humans are more important and their future should be like their past. These are fallacies – the past has gone, the damage has and will be done, and we are no more important than all the other species with which we share the planet. It is necessary to co-exist harmoniously with all the species in our communities, wherever and however we live, and those of us that live in the most modified environments, such as our cities, need to work harder than most.

*Top right: **The London Wetland Centre** at Barn Elms used to be good when it was a complex of old reservoirs, but now it's nothing short of brilliant. It is sure to go from strength to strength.*

*Right: Back on the riverbank – ratty has returned to town. **Water Voles** now 'plop' into the dykes and ditches at the London Wetland Centre.*

A FIELD GUIDE TO URBAN SPECIES

This field guide shows the flora and fauna that occur frequently in our urban and suburban environments and, therefore, are most likely to be encountered. These are the creatures and plants that profit from or put up with our chaos, whether it is on wasteland, canal sides, parks or gardens. Most are common, many familiar, some loved, others loathed, but an unbiased outline of the key characteristics and habits of each species is provided.

Mammals

Hedgehog *Erinaceus europaeus* 22-27cm (8⅔-10⅔in)
This nocturnal insectivore is unmistakable. The underbelly has a sparse, coarse fur and the male is generally larger than the female. With the use of a torch, it is easily located at night, especially in wet grassy areas. When disturbed it rolls into a ball protecting its legs, head and other extremities. Preferred habitat is grassy woodland edges, although rough ground in cities can also provide areas in which to search for food, a job aided by its acute hearing and sense of smell. Widely distributed in Britain, it is commonly found in gardens where it is useful in controlling slugs. Always be careful when burning garden waste in winter because hedgehogs are renowned for hibernating in compost heaps and unfortunately, many are cooked each autumn. Nevertheless, they are edible and I understand they taste rather like rabbit!

Common Shrew *Sorex araneus* 5-9cm (2-3½in)
Shrews are more frequently heard, when involved in their noisy territorial fights, than seen. During the summer, their shrill screams may last for up to 30 seconds. In winter and adult second summer pelage, the back of the Common Shrew is dark brown, its flanks paler brown. Its tail is bi-coloured and well haired. It can be found in almost every habitat, providing there is low cover, but is most abundant in thick grass, hedgerows or under bracken. In an urban environment, any patch of derelict wasteland, which has a thick growth of grass, will support a small population of this voracious insectivore. The Common Shrew is absent from Ireland, the Isle of Man and some of the Inner and Outer Hebrides.

Pygmy Shrew *Sorex minutus* 4-6cm (1⅔-2⅖in)
The Pygmy Shrew is uniformly coloured and is considerably lighter and smaller than the Common Shrew. It is medium brown above and has a dirty white ventral pelage, and the tail becomes naked in its second year. It has swift movements and a highly mobile snout. Found in the same habitats as the Common Shrew, the Pygmy Shrew occurs throughout the mainland.

Pipistrelle *Pipistrellus pipistrellus* body: 3.5-5cm (1⅖-2in); wingspan: 18-24cm (7-9½in)
The Pipistrelle is perhaps our most frequently encountered bat. Its upperparts are medium to dark brown, the underparts are similar but paler. Its ears are short, broad and extend nearly to the tip of its nose. It usually emerges about 20 minutes after sunset, and flies intermittently through the night. Flight is fast and jerky and like many other bat species, it hibernates and breeds in buildings. During the summer, large nursery colonies, of up to a thousand, are formed by females.

Grey Squirrel *Sciurus carolinensis* 23-30cm (9-11⅘in)
This 'Parkland Pirate' is easily distinguished from the rare and isolated Red Squirrel, which does not occur in any urban environment. The Grey Squirrel is larger and greyish. It has a less pleasing facial expression and its presence is advertised by scattered scales or stripped cores of pinecones, split shells and untidy, leafy dreys. It often utters a scolding "Chuck – charrey" or a low "Tuck-tuck", which can sound confusingly birdlike. Most abundant in mixed, mature hard woods throughout Britain, it also occurs in parks, large gardens and inner city areas, often becoming very tame, tolerating and even enjoying co-habitation with man.

Bank Vole *Clethrionomys glareolus* 9-12cm (3½-4¾in)
The Bank Vole, Field Vole, and Common and Pygmy Shrews are the most abundant small mammals on urban wasteland. The Bank Vole has a rich reddish-brown coat, prominent ears and a long tail. The presence of a vole is announced by its network of well-formed runways found at ground level, and blackish droppings. Abundant over all of the British Isles, the Bank Vole does not generally move far out into fields prefering the thicker cover of woodland hedgerow, dunes or scree with high herb and scrub layer. On wasteland, the breeding nests can often be found under discarded metal, cardboard, or wood.

Field Vole *Microtus agrestis* 8-13.5cm (3⅕-5⅓)
The Field Vole is a small greyish-brown animal with smaller ears, eyes and tail than the Bank Vole. Its field signs include piles of green-tinged, oval droppings. The Field Vole is most common in rough grassland, including young forestry plantations or any scrubby area where a lush growth of grass is present. It also inhabits wasteland. Only low-density populations are found in marginal habitats such as woodlands, scree or hedgerows.

Yellow-necked Mouse *Apodemus flavicollis* 7-12cm (2⅘-4¾in)
The Yellow-necked Mouse has large prominent ears and eyes and a long, nearly hairless tail. Its coat is a rich reddish-brown, with a yellow neck collar. It is predominantly nocturnal with peaks of activity at dusk and dawn. It rapidly scurries or leaps unless cautiously investigating. An intruder of suburban houses and expert raider of the larder, it is unlikely to be encountered outside the home. Its distribution throughout southern England and Wales is patchy and it is unlikely to stray deep into city centres.

House Mouse *Mus musculus* 7-10cm (2¾-4in)
The unwelcome presence of the House Mouse in shops, houses, factories, warehouses, farm buildings and refuse tips is often detectable by its concentration of faeces: small, dark, compact pellets. Its runways and footprints are discernible along regular routes through dusty places. A musty smell is also characteristic. It is a uniform greyish-brown above with relatively small eyes and ears. The tail is thicker and more scaly than our other mice and its sense of balance and abilities to run and climb enable it to intrude into all aspects of human life. Almost wholly nocturnal, it lives in a group with the potential to rapidly increase in numbers.

Common or **Brown Rat** *Rattus norvegicus* 20-29cm (7⅘-11⅖in)
The large size, pointed nose and long scaly tail preclude any confusion with any species of mouse, and its relatively small eyes and ears distinguish it from the rare Black or Ship Rat. It is typically found on farms, refuse tips or in sewers and warehouses, but is equally at home away from man's environment, around cereal and root crops and in hedgerows. Its coat is shaggy, grey-brown above. Found all over the British Isles, it can be detected by its burrows, which are often excavated under flat stones, logs or tree roots. Like mice and the Black Rat, its runs, found in buildings, show as dark, greasy smears on wood or brickwork. If a rat is found in town, it is certainly this species, and in some areas populations may be very dense.

Red Fox *Vulpes vulpes* 58-90cm (22⅘-35½in)
In the towns, clues to the fox's whereabouts are difficult to find. Footprints are easily distinguished from dogs because they are more oval in shape and the two centrally placed pads do not overlap with the two side pads. Its faeces are usually pointed, black with a characteristic odour and, when fresh, may be linked by hairs. Frequent calls – high-pitched barks, and a single coarse wailing bark – can be heard in winter. Foxes occur all over the British Isles and can be seen in all of our major cities.

Badger *Meles meles* 68.5-80cm (27-31½in)
This nocturnal omnivore has short legs, with well-clawed feet for powerful digging. Distinct footprints are visible in the sand outside their setts. The male has a broader head and a whiter, more pointed tail than the female. Setts, often hidden in the dense cover of wasteland or overgrown gardens, are distinguished from Fox earths by the diameter of their burrows, which must be at least 20cm (7⅘in). The Badger has acute senses of smell and hearing, but its eyesight seems poor. It occurs in every county in Britain and can frequently be found in suburban and urban environments foraging for its diet of Earthworms.

Birds

Grey Heron *Ardea cinerea*
94-100cm (37-39⅖in)

This is the largest common bird in the British Isles. A long dagger-shaped, yellow bill and black flimsy crest make this bird familiar. Its flight is powerful yet slow, with deep wing beats, and it has a heavy, rounded flight silhouette, with wings bowed and legs trailing. Its voice is a deep, harsh 'Kraarnk'. It frequents water meadows, rivers, lakes and seashores where it stands motionless for long periods holding its long neck primed for stabbing at prey. It also visits urban lakes and riversides, occasionally scooping up goldfish from garden fishponds.

Canada Goose *Branta canadensis* 90-100cm (35½-39⅖in)

Introduced from North America, it is now Europe's largest free flying goose. Its black neck and head and white face patch, which is coffee stained in young birds, are unmistakable. The honking call, uttered in flight and on the water, is a loud, double syllable and nasal 'aww-lut'. It will accept any waterway as a home and is rarely shy, frequently bullying all the other waterfowl in its way. This habit along with its grassy poo, which litters park lawns, make it less than popular.

Mallard *Anas platyrhynchos* 60cm (23⅔in)

Mallards are most widespread of the British wildfowl, and are frequently seen on ponds in parks and urban waterways and even small patches of water in car park forecourts. The males have glossy green heads, narrow white collars, and curled back centre tail feathers. Females are mottled brown, showing a double white wing bar in flight. Believe it or not their call is a deep 'Quack–quack–quack'! I suppose most will also answer to the name of Donald or Daffy as you throw them your sandwiches!

Sparrowhawk *Accipiter nisus* 28-41cm (11-16in)

This species has short rounded wings, a long tail and rapid wing beats between long glides in flight. The female is much larger than the male. Its call is a rapid 'Kek-Kek-Kek' and is conspicuous in spring when performing its territorial and mating display over woodlands. Increasingly common in suburban parks and commons, Sparrowhawks regularly forage over gardens for their diet of small birds. There are frequent reports of them crashing into conservatory-type windows due to pursuing their prey to the very limit of their capabilities.

Kestrel *Falco tinnunculus* 33-39cm (13-15⅖in)

This Kestrel is often seen hunting over patches of wasteland, the sides of motorways and amongst the hubbub of suburbia. It is distinguished in flight by its pointed wings, long narrow tail and hovering behaviour. Their flight consists of rapid wing beats and occasional short glides. Male Kestrels have a lavender blue-grey tail with a thick black band. Females have a barred chestnut tail. Their call is a shrill repeated 'Kee, kee, kee, kee'. They occur all over the British Isles and are famous for their inner city nest sites, which include high rise window boxes, electricity pylons, office blocks and the more traditional cathedrals.

Peregrine *Falco peregrinus* 38-51cm (15-20in)

At rest the broad chest, dark head and thick black moustache can be distinctive, whilst in the air the adults are slate-grey above and pale beneath, and finely barred on the lower chest and wings. Flight is purposeful with quick shallow wing beats. Outside the breeding season Peregrines are pretty quiet but a scolding cackle is used as an alarm call. These birds are enjoying a real urban spread so keep your eyes peeled and ears open, especially in summer when their young are conspicuous and noisy.

Moorhen
Gallinula chloropus
33cm (13in)

Smaller than the Coot, the Moorhen has an irregular white streak along its flanks and conspicuous white undertail coverts. It has a deliberate probing gait and the tail is constantly flicked. Its voice consists of a harsh 'Kr–r–rk' and an array of explosive squeaking notes. The Moorhen's flight is low and weak with legs dangling, and it rises noisily from the water by pattering madly over the surface. A common bird of waterside margins, it can become quite tame where not molested, often occurring on city centre ponds.

Coot *Fulica atra* 38cm (15in)

This robust bird has a jet-black head with a conspicuous white frontal shield on its forehead. The Coot walks and runs well and frequently forages away from water. Flight is weak and it lands on water uncomfortably and noisily. It is gregarious in winter when huge numbers build up on reservoirs. Larger areas of water are preferred, although its nest is situated amongst reeds and other aquatic vegetation. It is quite common in our inner cities.

Black-headed Gull *Larus ridibundus*
36-40cm (14⅖-15¾in)

Smallest of the commonly seen gulls, it has a slender build and pointed wings. Its reddish-orange legs and red bill are found on no other British species, and in summer, adults are marked with a thick, chocolate-brown hood. In winter this dissolves to a dusky spot behind the eye. Its flight is lighter and more effortless than our larger gulls and it occurs on arable fields, sewage farms, and refuse tips outside city boundaries. Its raucous voice is gull-like. Although easily lured by your offered titbits, it remains a wary visitor to our towns.

Feral Pigeon *Columba livia* 30-35cm (11⅘-13⅘in)
If you cannot recognise this bird I demand you drive straight to the sea and cast your binoculars into the waves! A descendant of the Rock Dove, it appears exactly the same in size and shape but is of a motley colouring. Any mix of white, grey, reddish-grey, black, brown of shiny green-and-purple decorates this mongrel, which flocks in our towns and cities. Widely despised and discouraged this pigeon is a dogged survivor, even when faced with the most derelict environments, and ranks as one of the most successful species of urban habitats.

Woodpigeon
Columba palumbus
40cm (15⅘in) 
This is the largest of our pigeons. It has a conspicuous, broad, white wing band in flight and is marked with a glossy, green-purple-and-white smudge on its neck. Its display flight is similar to that of the Feral Pigeon, but it is altogether a larger, fatter bird, and is resident throughout the British Isles. In towns and cities it will frequently be solitary and mingle freely with town pigeons, but will only alight in the larger gardens or parks because of its wary nature.

Collared Dove *Streptopelia decaocto* 28-32cm (11-12⅔in)
This very pale, long-tailed and broad-winged Dove has a deliberate flicking flight and is frequently seen buzzing over gardens and parks. It is told from its country cousin, the increasingly rare Turtle Dove *Streptopelia turtur*, by its pale dusty brown upperparts, and the narrow black collar on the back of its neck. In flight the white terminal half of its black tail is diagnostic, and its call is an irritatingly deep 'Coo–cooo, coo'. It nests in trees and shows an affinity for man's habitation, having spread from its first nesting site in 1955 to be widely distributed throughout the British Isles.

Tawny Owl *Strix aluco* 38cm (15in)

This nocturnal owl has a heavy build, large black eyes and lacks any ear tufts. It varies in colour from a warm brown to greyish-brown and is marked with bold dark streaks. Its familiar call is a shrill 'Ke-wick' and its song a very characteristic 'Hoo-Hoo-Hoo'. Tawny Owls are widely distributed across the British Isles, and frequent mature mixed woodland often occurring in parks and gardens even deep into our city centres. Basically you ought to know one if it hits you in the forehead!

Swift *Apus apus* 16cm (6⅓in)

Swifts are distinguished from the Swallow family by their long scythe-shaped wings and sooty black plumage. A whitish chin is invisible as they ricochet in their mad screaming groups around rooftops. Their voice consists of a prolonged, shrill and piercing screech. They are entirely aerial birds and only perch at their nest site. Situated under the eves of buildings a scant nest of grass, which is snatched from the air, contains their two plain white eggs. Swifts appear from late April and depart at the end of August and can be found nesting in buildings across the British Isles.

Rose-ringed Parakeet *Psittacula krameri* 28-45cm (11-17¾in)

This bright green and extremely noisy exotic will not go unnoticed. A native of Asia and introduced to the UK, it is prospering in our run of mild winters. It is particularly abundant in Kent, Surrey and South London, but has spread widely and can often be seen in large rowdy flocks. Bird tables are plundered, orchards are ravaged and parks with mature trees provide nesting grounds. Fast flight, dark underwings and a very long and narrow tail are characteristic, as are the red bill and the male's rose neck-ring.

Great Spotted Woodpecker *Dendrocopos major* 23cm (9in)

The Great Spotted is distinguished from the much smaller Lesser Spotted by its black back, large white shoulder patches and crimson undertail coverts. The underparts are white with a sharply defined red patch below the tail. It produces a frequent and sharp 'Tchick' or 'Kik'. The Great Spotted is a frequent visitor to bird tables and supplies of fat and can be found all over England extending into Scotland and Wales.

Lesser Spotted Woodpecker *Dendrocopos minor* 14.5cm (5¾in)

The smallest woodpecker in Europe, the Lesser Spotted is distinguished from all others by its sparrow size, closely barred back, white underparts and an absence of any red on the tail coverts. The male has a dull red crown whilst the female's is whitish. In a natural situation the Lesser spends most of its time in the upper branches and is only an infrequent visitor to gardens because it is less common than the Great Spotted Woodpecker.

House Martin *Delichon urbica* 12.5cm (5in)

The House Martin is the only European Hirundine with a pure white rump. Its underparts are also white, whilst its head, back, wings and tail are an iridescent blue-black. The tail is short and forked, lacking streamers and their bill is small, although they have a large gape to facilitate their aerial feeding habits. Smaller than the Swallow, it often flies higher when feeding. Its song is a weak, pleasant chirruping twitter and it is frequently found cupping its enclosed mud nests under the eves of houses and barns over most of the British Isles. Summer visitors, they arrive from mid-April and remain until mid-October.

Pied Wagtail *Motacilla alba yarrellii* 17.5cm (6⅘in)

In summer, male Pied Wagtails are almost black backed, but retain a double white wing bar and white outer feathers in their long and conspicuous tail. Individual birds vary greatly in their plumage, but their bobbing and wagging tail separate this species from others. They have an undulating flight and a lively 'Tchizzik' alarm call but a talentless, twittering song. Natural and unnatural cavities, such as holes in walls, sheds or pipes, generally situated near water, provide nesting sites. Large winter aggregations can occur in city centre trees, on large buildings or on sewage farms where they find an abundance of their insect food.

Wren *Troglodytes troglodytes*

9-10cm (3½-4in)

A tiny, hyper-active, mouse-like bird, it has a cocked tail and short neck that give it a ball-like appearance. A rattling 'churr' or long, liquid, canary-like song, generated from cover, belies the size of this bird. It needs dense undergrowth or scrub but is still regular in heavily vegetated gardens where it ferrets in pursuit of insects. Males make well-hidden, dome-shaped, moss nests for females to choose and line before laying a clutch of up to eight eggs. The colder it gets the bolder they become and the more they need you. Mealworms provide a winter treat – buy some!

Robin *Erithacus rubecula* 14cm (5½in)

This is the plump, neckless, red-breasted bird that has escaped from the cover of nearly every Christmas card. Adults of both sexes have rich orange breasts and forehead and uniform olive-brown upperparts. The varied warbling phrases of their song can be heard all year round. In the British Isles

Robins are bold, tame birds often associating and confiding in man. They forage in gardens, hedges, coppices and woods amongst the undergrowth searching for their mixed diet of berries and insects.

Song Thrush *Turdus philomelos* 23cm (9in)

This brown-backed bird has creamy-yellow, heavily spotted underparts and a clean orange underwing. It is the typical garden thrush whose feeding behaviour consists of hops interspersed with short runs, punctuated by upright watchful pauses, often cocking its head to listen intently for its prey. The voice is a loud 'Tchuck' and song a musical collection of repeated phrases delivered from a perch. It is common throughout the British Isles and famous for its snail-smashing exploits, clean mud bowl nest, and beautiful blue and black spotted eggs.

Blackbird *Turdus merula* 23-29cm (9-11⅖in)

On clear spring evenings the song of the Blackbird rings throughout suburban Britain. It is a melodious warbling, distinguished from the Song Thrush's by its purer fluty notes and lack of repetition. It is terminated by a collapse into a weak and unmusical riot of notes, and is delivered from treetops and television aerials. The rattling alarm call can be heard whenever a cat is on the prowl. The male is black with a bright orange bill, whilst the female has a uniform dark brown plumage. It is common in the gardens and parks of all of our cities.

Blackcap *Sylvia atricapilla* 14cm (5½in)
This elegant warbler has longish wings and tail and an attractive and lively attitude. The male has a glossy black crown whilst the female has a red-brown crown. In its swift, jerky flight the Blackcap's tail is conspicuously long. Its voice is an emphatic 'Tack, tack' repeated rapidly when alarmed and its song is a rich warbling, which is more varied and less

sustained than the Garden Warbler. During the winter, it appears on bird tables. Blackcaps are not strictly garden birds but the recent run of wet and warm winters has led to more of them remaining residents instead of migrating south.

Great Tit *Parus major* 14cm (5½in)
The largest of the British tit species, it is the least acrobatic and the most unruly. It is active, inquisitive and always quarrelsome, often routing other visitors from bird tables, whilst producing a scalding Blue Tit-like 'Chi–chi–chi'. It has a glossy black head and neck, white cheeks and yellow underparts. This species also inhabits nest boxes and is an equally common visitor to garden food supplies, often foraging on the ground as well as on suspended nut bags.

Coal Tit *Parus ater* 11.5cm (4½in)
Slightly smaller than the Blue Tit, this is the only black crowned tit with a bold white patch on its nape. It has a disproportionately large head, two white wing bars and a forked tail. It is always restless and acrobatic and emits a thin clear 'Tsui' and 'Susi-susi' call. The Coal Tit's song is a repeated clear 'Seetoo', which is

more rapid and less vigorous than the similar notes produced by the Great Tit. It is common throughout the British Isles where it is fond of conifers, but often visits gardens and nut bags particularly during the winter.

Blue Tit *Parus caeruleus* 11.5cm (4½in)
The most common British tit, it is the only one with a bright cobalt blue crown, wings and tail. It has a small compact crown, often raised when agitated and its

flight is fast and undulating with rapidly flicking wingbeats. It has varied call notes based on a 'Tsee–tsee–tsee–tsit', a phrase, which is used to pre-empt its trilling song. A regular visitor to bird tables, it is easily coaxed into breeding by a well-made and well-placed nesting box.

Magpie *Pica pica* 46cm (18in)
This bird should be familiar to everyone regardless of any ornithological interest. Even its rapid 'Chak–chak–chak–chak–chak' call should be well known. Magpies are wary birds who often build domed nests in the tops of hawthorns. Within suburbia, nearly all types of tree are requisitioned for nesting, and recently they have begun to use man-made structures to support their nests. They are increasingly common in England and in some places so enjoy human habitation that they have become very common attracting alot of unwelcome and unfounded blame for their occasional habit of robbing smaller birds of their eggs or young.

Jackdaw *Corvus monedula* 33 cm (13in)

The smallest of our crows, these naughty, perky and inquisitive birds are often seen larking about on our rooftops. Jackdaws are highly gregarious, often mixing with Rooks and Starlings, but are distinguished from other crows by their grey nape and ear coverts, smaller size and shorter bill. Their call is an unmistakable 'Chak' and, when excited, a wild 'Chaka–chaka–chak'. Generally distributed throughout Britain, they are commonly found in association with man's habitation.

Carrion Crow *Corvus c. corone* 44-50cm (17⅓-19⅔in)

A clever and adaptable bird, which due to an omnivorous diet can thrive almost anywhere. Entirely black with a greenish-blue sheen in good light, it is easily told from the Jackdaw through its larger size and lazier, listless flight. A loud 'Kraa–Kraa–Kraa' call is typical and crows often hang out in pairs or families rather than in flocks. Ever cautious, the crow is incredibly difficult to approach, which is a real shame as it is the most intelligent of all birds.

Starling *Sturnus vulgaris* 22cm (8⅔in)

This noisy, squabbling bird is widespread throughout the British Isles in all our urban and agricultural habitats. Its flight silhouette shows pointed wings and fan-shaped tails. It is glossed black with bronze-green and purple. In winter, its plumage is closely specked with cream and at rest it often appears decidedly obese. It frequently mimics other species or man-made sounds and its song is a harsh rambling medley of whistles, clicks, rattles and chuckles delivered from television aerials, chimney pots or washing lines. Very gregarious, it feeds, roosts and nests in noisy throngs in city centres and suburbs.

House Sparrow *Passer domesticus* 14.5cm (5⅔in)

This cheeky, aggressive, garrulous bird almost always nests in buildings, often in colonies, and occasionally bushy woods and roadside trees. The male has a dark grey crown and black throat. The female and juvenile are dull brown above. There is no real song, only a mixture of grating, twittering, and chirping notes, and a loud 'Cheep'. A serious and rapid decline has diminished the south's Sparrow populations but in some places in the north they seem okay.

Chaffinch *Fringilla coelebs* 15cm (6in)

Britain's commonest finch, it is often seen hopping in a series of low, jerky bounds with tail down as it feeds on the ground. Its flight is extremely undulating and displays a green rump and double white wing bars. The male is pinkish brown below with a slate-blue crown and nape whilst the female is olive-brown above and lighter below. Its song is an extrusion of notes terminating in a 'Choo–ee–o' and the call is a loud repeated 'Chwink'. Frequently found in gardens, parks and commons, it produces a neat mossy nest and four or five beautifully marked reddish eggs.

Greenfinch *Carduelis chloris* 15cm (6in)

A Canary-like twittering with mixed call notes is delivered from bush tops or in a rather bat-like song flight. Males are olive-green with a conspicuous yellow rump, whilst the females are more brown overall, both sexes have short forked tails. They are quarrelsome on bird tables and strongly undulating in flight. In winter, they flock with other finches, but during summer they frequent gardens nesting in hedges, bushes and small trees. They are distributed throughout the British Isles and are resident in most suburban bird communities.

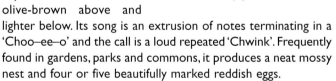

Reptiles and Amphibians

Common or Smooth Newt
Triturus vulgaris
Up to 11cm (4⅓in)

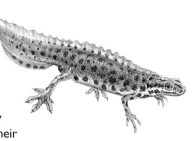

This newt often has a spotted chin! Breeding females are yellow-ish-brown above with small dark spots, which may fuse into two lines along their backs, and a vague stripe on the side of their heads. Their belly is marked with dark spots over a bright orange-yellow, or red base colour. Breeding males develop a prominent crest, which runs the length of the body and tail and are marked with characteristic large dark spots. It is widespread, found in any clean freshwater ditch, pond, stream and drain even within the city centres.

Common Toad *Bufo bufo* Up to 15cm (6in)
This is the largest European Toad. It has a very warty skin and prominent swollen glands behind its eyes. Usually brownish, it varies from sandy to almost brick red, dark brown, greyish or greenish. Their eyes are deep gold or copper coloured. At night males produce a weak, high pitched, rough 'Qwark–qwark–qwark'. Always nocturnal, they can be found in a wide variety of habitats, and in spring, hundreds or even thousands of toads appear in the open, migrating to breeding pools. At this time, 'Toad Crossings' are a part of life for many school children who act as 'Lollypop Men' for these amphibians ensuring their safety as they cross roads to reach their soggy destinations.

Common Frog *Rana temporaria* Up to 10cm (4in)
This is the commonest species of frog over much of Europe. It has short hind limbs and is marked with a dark and distinct eardrum. The colour is extremely variable, ranging from grey, brown, pink to olive and yellow, often with an inverted V-shaped mark between its shoulders. Their skin is much smoother and less warty than a toad's. A dull rasping 'Grook–grook–grool' is produced during the breeding season when this species migrates to traditional ponds, often very early in the year, to spawn. Today urban ponds are becoming an essential stronghold for this species which is rapidly declining in rural areas.

Common or Viviparous Lizard *Lacerta vivipara* Up to 18cm (7in)
This pretty little lizard is a long-bodied, short-legged creature with a small, rather rounded head and thick neck and tail. Its colours are very variable, but it is generally brown with dark sides, a vertebral stripe and a number of light streaks. The young, which appear at the end of the summer are almost black.
Common Lizards frequent rough ground, field edges, heaths, bogs, and grass-land, but can often be found on urban wasteland where they favour the warmth provided by discarded metal, cardboard or plastic.

Slow Worm *Anguis fragilis* Up to 50cm (19¾in)
Slow Worms are very smooth-scaled, snake-like lizards, usually brown or grey, very occasionally marked with blue spots. Females often have a central dark stripe, dark sides and belly whilst males are more uniform. They are usually slow moving and likely to be seen during the evening or after rain. They can be found on scrub, heath, railway embankments, gardens, hedgebanks and wasteland, wherever there is extensive ground cover. They are often found lying beneath old iron cardboard and flat stones. If you do search under such debris please put it back because many other animals also rely on it for shelter.

Grass Snake *Natrix natrix* Up to 120cm (47⅓in)
This beautiful snake is decreasing in the British Isles. It is nearly always found in association with water, so ponds, ditches and drains are its usual habitats in the urban environment. Its body colour is very variable, but all specimens have a yellowish-white, black-bordered collar just behind the head, and the eyes are large and yellow. A diurnal reptile, it swims well and feeds on frogs, toads, newts, tadpoles and fish. It is quite harmless, and if disturbed will rarely bite, choosing instead to either feign death or emit the evil smelling contents of its anal glands. A scent that will linger on your hands.

Insects

Silverfish *Lepisma saccharina*
13mm (½in)

The Silverfish has a tapering, carrot-shaped body coated with shiny scales, long thread-like antennae and three long segmented tails. This primitive insect has simple, biting mouth-parts and is a frequent inhabitant of food cupboards, where it feeds on scraps of paper, cartons, glue and spilled flour. If you catch a Silverfish, it normally slips away leaving a coating of silver on your fingers. In the past, it was regarded as a pest of books because of its taste for bookbinding glue. Unfortunately, the Silverfish won't enjoy the glue used in the production of this volume, so this book is likely to linger on your shelf!

Broad-bodied Chaser *Libellula depressa* 40mm (1½in)

There is nothing depressing about these little pond-trotting jewels, or at least not the males who sport a blue abdomen. The

dowdy females display the same aerial finesse, but are yellow-brown. Virtually any still or slow-flowing water will attract these highly territorial predators who chase off virtually anything insect and airborne that strays into their space. They range as far north as Cumbria in the UK. True survivors, they'll even take to brackish waters and they're always entertaining – if you like death in the air!

Common or Oriental Cockroach *Blatta orientalis*
22mm (⅚in)

This pear-shaped insect is by far the most likely cockroach to confront you in your cupboard. The females are black-brown with wings reduced to tiny flaps whilst the male is reddish-brown with wings that nearly reach to the end of the abdomen, but he remains flightless. It has long and always quivering antennae, long, spiky legs, and a compact, flattened body, and can be found in bakeries, restaurants, sewers, and houses.

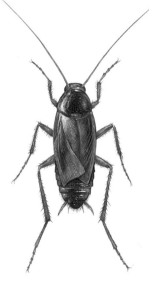

Wall Brown Butterfly
Lasiommata megera
45mm (1¾in)

A once common species, the Wall Brown, which is apparently decreasing, emerges in May and a second brood is on the wing from late July through to early September. The male is a bright tawny-brown colour marked with a thick band of scent cells across the forewing. In contrast, the slightly larger female has a lighter yellowish-brown ground colour, with a more prominent hindwing spot. In flight, this quick and restless butterfly frequently lands on bare ground. It can be found on wild wasteland and commons as far north as southern Scotland.

Speckled Wood *Pararge aegeria* 45mm (1¾in)

This species frequents woodland glades, pathways and shaded country lanes, but can be found well within the city boundaries in any overgrown area of parkland, wasteland and large gardens.

It has a weak flitting flight and often alights in sunspots in the shadow of overhanging trees. Here it sits with wings wide open sunning itself, showing its characteristic black-brown coloured wings, marked with 11 yellowish blotches. The slightly larger female has a lighter ground colour. The adult lives between 12 to 14 days and is on the wing from April through to late June; second broods often appear in August and in a dry summer remain on the wing as late as October.

Meadow Brown
Naniola jurtina
50mm (2in)

Meadow Browns are dark brown with white, centred eyespots near the tip of each forewing. They can be ridiculously common in some habitats, occurring in sub-plague proportions over thick patches of thistles and brambles. Their slow floppy flight and apparent loathing to leave their resting places, make them easy to observe and photograph. They frequent meadows, commons and heaths, but are equally happy on patches of wasteland, however untidy, within the city boundaries. They are on the wing from July until late September and occur over nearly all of the British Isles.

Red Admiral *Vanessa atalanta* 60mm (2⅖in)
This famous butterfly should need little introduction. The ground colour of the similar sexes is a rich velvety black, marked with wide, scarlet bands across the forewings and hindwings, the edges of which are decorated with blotches and spots of white. Its strong powerful flight wends it around Buddleia, Valerian, rotting fruit and a bevy of cultivated flowers. Individual Red Admirals will return daily to their favourite bushes. This migrant is particularly common in late summer when it visits gardens, parks, waste ground and nettle beds.

Small Tortoiseshell *Aglais urticae* 50mm (2in)
This is probably our most common brightly coloured butterfly. Its flight is jerky and often Tortoishells chase each other around their nectar sources. It feeds with wings open but roosts with them closed, revealing its dull and drab undersides. In contrast their uppersides are a rich tawny-orange with a row of violet-blue spots encircling all four wings near their outer margins. At the front edge of the forewings are three well-defined black blotches, followed by a yellow one nearer to the body. The first broods emerge in June and can be found on the wing until September, in parks, gardens and flower-filled wasteland, almost everywhere throughout the British Isles.

Peacock *Inachis io* 60mm (2⅖in)
Like the Small Tortoiseshell, this species is common all over the British Isles although its numbers fluctuate from year to year. It can be found in gardens, parks and on wasteland wherever nettle beds provide a food source for its glossy black and spiny caterpillars. The Peacock's ground colour is a rich wine red and conspicuous 'peacock' eyespots dominate each forewing. A dark purplish-brown underside makes it very difficult to see because it hibernates upside down in dry out-buildings. If disturbed during hibernation it may produce a threatening rustling noise by rubbing its hindwings together.

Comma *Polygonia c-album* 45mm (1¾in)
The upperwings of this pretty butterfly are rich reddish-brown. The undersides of both sexes have a bold white C-shaped mark in the centre of each hindwing, which gives rise to its name. Its most distinctive feature, however, is the irregular serrated edge of its wings, which superficially gives this butterfly a 'damaged' appearance. Unfortunately, the Comma seems to be decreasing over much of its range. Like the Peacock it hibernates over winter to re-appear in late March. Adults appear in July and August and Commas can be found on the wing until October. Their favourite haunts are woodland edges and clearings or anywhere where bramble and rotting fruit appear.

Common Blue *Polyommatus Icarus* 32mm (1⅛in)
This is our most common blue butterfly. The sexes are quite distinct. The male is a clear violet blue and the female a wholly dull brown. The underside of the male is pale bluish-grey and the female pale coffee. It frequents hills and meadows throughout the British Isles and any rough pasture or wasteland can provide a home for this insect. It has a quick fluttering flight and often alights on grass stems to sun itself with its wings half open. There are two broods, the first emerging in late May and throughout June, the second appearing in August and September.

Common Clothes Moth *Tineola visselliella* 10mm (½in)
This tiny, tatty moth often has a silvery or golden sheen on its wings, and without the use of Napthalene can do considerable damage to all keratin-containing materials, such as hair, wool, silk and feathers as well as cotton and stored cereal products. Its larvae cause the damage by constructing small tubular cases from fragments of the material they have destroyed. Although the damage caused by the Common Clothes Moth has decreased in recent years due to the availability of effective insecticides, millions of pounds are still spent annually in Britain on its control.

Common Crane Fly

Tipula paludosa wingspan
15-65mm (⅔-2½in)
In their adult form these 'Daddy-long-legs' are frequently attracted to our houses by the lights. Universally loathed and yet entirely harmless, it is the larvae of these insects, which damage crops by eating their roots. The winged adults only lap up liquid at the end of a beak-like head extension, and their greyish body and legs seem extremely fragile,

often breaking off when attempts are made to dissuade these flies from our company.

House Fly *Musca domestica* 6mm (⅕in)

The House Fly is the commonest of the flies to buzz around our buildings. It is one of the most widely spread of all animals,

because it has followed man all over the world. It has a black thorax and a yellowish-brown abdomen often possessing honey coloured legs. As with all true flies, its hindwings are reduced to tiny balancing organs and its habit of contaminating our food and laying eggs in decaying material makes it a highly persecuted and unpopular insect. I once had a pet House Fly, Arnold, who spent over a week in my bedroom feeding on tasty leftovers I provided. He was good company: quiet, understanding, yet witty and lively, and as yet I have not died of any dangerous diseases!

Common Wasp *Vespula vulgaris* 20mm (⅘in)

This insect requires no description, and no sane person should fear it, ever. Common Wasps build large nests in our homes or in natural cavities where up to 4,000 workers may live and thrive, each feeding on many harmful insect pests. Indeed, wasps are entirely harmless until you are mean with your sweets or beer and begin to behave like an air traffic controller on a bad day. Wasp workers should be enticed by providing a little sweet food to supplement their insectivorous diet and then examined at close quarters. Their lifestyle is, after all, far more ordered and socially successful than our own! Thank goodness most of us do not have stings!

Garden Bumble Bee

Bombus hortorum
10-40mm (⅖-1½in)
There are many species of bumble bee and their common name often relates to their hairy warning colouration, Red-tailed, Buff-tailed, etc.

They are all social insects, which make their nests underground, and are always common in gardens where they collect nectar and pollen. A bumble bee key was produced by the WATCH organization a few years ago and this provides an excellent way of separating the common species which we find in our gardens and wasteland across the country. Most are large, hairy, slow moving and quite harmless and should be worthy of only our constructive observation.

Devil's Coach Horse *Staphylinus olens*
25mm (1in)

This furiously fast and bizarre-looking beetle sends tingles down the spines of children with its relatively large mandibles. It appears odd because its wing cases are much reduced leaving its soft abdomen exposed. This also allows its tail to be raised over its head in an aggressive/defensive posture. There is no sting nor earwig-type pincers. A 'Rove' beetle, it is a predator, nocturnal, typically found under logs or litter in the daytime and seems imbued with inexhaustible energy. Cup one in your hands and it will run and run and run. Stopping occasionally to try and nip you.

Violet Ground Beetle

Carabus violaceus
40mm (1½in)
There are over 350 British species of Carabid, more commonly known as Ground Beetles, each a predator capable of rapid running. Many are nocturnal and can only be found in daylight hiding under logs, stones and other discarded debris. Both adults and larvae are carnivorous, actively hunting for worms, slugs and other insect prey. Most are black but the Violet Ground Beetle is a beautiful, rich dark purple and can often be found lying helplessly on its back in daytime. This sorry condition is usually due to a bout of fighting the previous night or the garden birds' distaste of its flesh. Attempts to right them are futile because they seem to lose all sense of co-ordination.

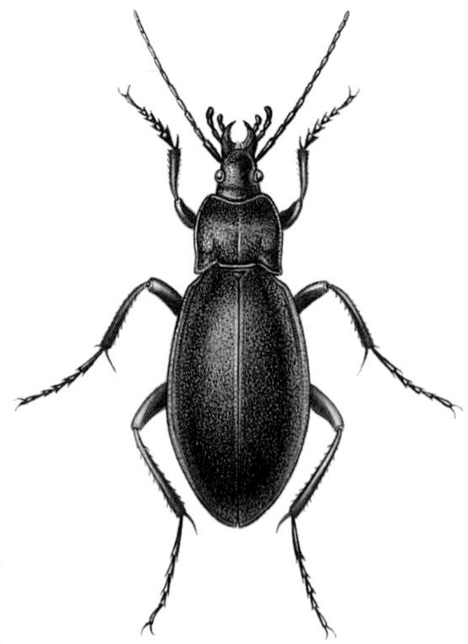

Seven-spot Ladybird
Coccinella 7-punctata
7mm (⅓in)

Just one of Britain's 45 species of ladybirds, the Seven-spot Ladybird has a bright red elytra, with, as the name suggests, seven black spots. Ladybirds are predatory and their larvae control vast numbers of Green Fly and other garden pests. They come in a range of bright colours, generally black, red and yellow; colours that advertise to birds their bitter taste. When handled, ladybirds often ooze drops of pungent blood, which stain your fingers yellow. The many variable species of these beetles are best separated by the use of a good field guide or the guide produced by the WATCH organization (*see* page 141) a few years ago.

Woodworm or Furniture
Beetle *Anobium punctatum*
4mm (⅖in)

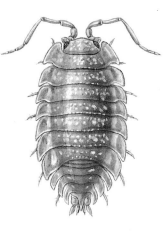

This notorious pest is a small, compact, slightly elongated, choc-olate-brown beetle whose larvae tunnel into dead wood. It is in the rafters and furniture, which this spe-cies can quickly reduce to dust, that it attains pest status. Well-polished furniture is safe from attack but adults can lay their eggs in any small crevice and it is not until they emerge leaving their wormholes that we belatedly learn of their presence. Fortunately, for collectors of Chippendale *et al.* they are easily controlled with a range of insecticides. Fortunately, for Woodworm, there are plenty of old barns, sheds and beds that go untreated.

Common Woodlouse
Oniscus asellus 15mm (⅔in)

The Common Woodlouse with yellow spots and tan-marked grey carapace, is one of more than 40 spe-cies in the UK. All are tank-like with seven pairs of legs, eyes, antennae, chewing mouth-parts and modified rear limbs used for respiring and mating. Some species such as the Pill Woodlouse, *Armadillidium vulgare,* can roll into a perfect ball or 'pill'. Look out for the Common Woodlouse, along with the all-grey *Porcellio scaber,* under your doorstep or anywhere dark and damp where a light salad of decaying vegetation will sustain a small array of these harmless crustaceans.

House Spider *Tegenaria gigantea*
20mm (⅘in)

This species is one of three easily confused House Spiders, which can be found throughout most of Britain. In the past, its habitat was under logs, holes in banks and behind bark, but, today, it is more often found in houses, garages and gardens. The spider is brown and has two rows of darker radiating marks running down its prosoma, whilst its bulbous opisoma is a yellow-brown. The legs are long, twice the body length in females, and three times the body length in males, giving the spider the ability to move rapidly along your skirting boards. Perfectly harmless, it should not be killed on sight. If you object to it sharing your bathroom, simply throw it out of the window.

Garden Spiders *Araneus diadematus*
12mm (½in)

Common in gardens where it spins its large and characteristic web, the Garden Spider has a large bulbous abdomen. By the end of the summer when it has moulted several times, it appears huge and imposing while resting in the middle of its silks. It is very varied in colour, ranging from black through to ginger, but is always marked with rows of white spots and transverse, white blotches. Like all British spiders, this one is perfectly harmless and should be left in place and peace to reduce the number of flies about your dwelling.

Brown-lipped Banded Snail *Cepaea nemoralis* 30mm (1⅖in)
One of two species of clearly banded snail occurring in urban Britain, it favours grassy areas of scrub. What's interesting is its polymorphic variation – the shell's range of colours – which provides it with the most effective camouflage in any given environment. Pale pink or yellow-based snails are whorled with various numbers and widths of rich reddish bands, which effectively darken the shell. The relative abun-dance depends on what is eating it and where. Song Thrushes love them.

Plant Life

Traveller's Joy *Clematis vitalba* Up to 30m (32⅖yd)

Often known as 'Old Man's Beard', this woody climber produces greenish-white flowers between July and September and woolly greyish-white plumes, reminiscent of balls of dirty wool, throughout the winter. The leaves are opposite pinate and have twining stalks. Traveller's Joy is supported by scrub, hedgerows, wood margins, fences and wasteland and often this voracious perennial can completely smother other shrubs, particularly Elder bushes *Sambucus nigra*. It is a widespread plant of lowland Britain and can be found south and east of a line between the Severn and Humber.

Meadow Buttercup

Ranunculus acris Up to 50cm (20in)

This is one of three common species of British buttercup, which can be found in our damp meadows and pastures. The flowers of the Meadow Buttercup appear from May onwards and it differs from the Creeping Buttercup *R. repens* in that it has no creeping runners or short leaflets. The third common species, the Bulbous Buttercup *R. bulbosus* occurs on drier and better drained soils and is particularly fond of chalk and limestone grassland. It is the earliest of the three to flower and can be told from its cogeners by its thick, swollen, corn-like base to stem. All the species have conspicuous bright yellow and glossy flowers, which often appear carpeting fields in yellow during May.

Broom *Sarothamnus scoparius* Up to 2m (79in)

Across the British Isles, Broom can be found on the well-drained, acid soils of rough grassland, heaths, roadside verges and wasteland. It flowers in May and June and is pollinated by bumble bees. The large, bright golden yellow, leguminous flowers often provide an attractive sight in derelict areas. Superficially, this shrub appears like a spineless Gorse but it has markedly angled stems and small lanceolate leaves. It often develops in dense, untidy and impenetrable clumps around the edges of car parks or road verges, even in city centres.

Bramble or Blackberry

Rubus fruticosus Up to 4m (157in)

This very familiar, very variable, totally clambering and evergreen perennial is famous for its tasty fruits, which turn from green through red to purplish-black and fill tarts, pies and jam pots across the land. The flowers, which appear from May to September, are delicate, white or pink, five-petal structures, which are pollinated by a variety of insects. The stem is prickly as are the pinate five-toothed leaflets, which turn reddish-purple in autumn. Some 400 'micro species' have been described in the British Isles where this plant is widespread, and can be found in woods, hedges, and waste places as well as the top of heaths and cliffs.

Rosebay Willowherb *Chaemerion angustifolium* Up to 1.5m (59in)

This vigorous and showy perennial has narrow lanceolate leaves and loose, tapering spikes of large, pinkish-purple flowers, which have unequal petals and prominent, reddish sepals. Seeds are exposed when its narrow fruits peel and clouds of untidy down then coat the vegetation until they are dispersed by the wind. One of a host of similar willowherb species, which can be found across the British Isles, it favours fire sites, felled woodlands, derelict buildings, waste grounds and sand dunes, often forming dense spreads on recently disturbed soil.

Ivy *Hedera helix* Up to 30m (32⅖yd)

This evergreen and woody climber can be found creeping all over the British Isles. Its flowers appear from September to November and are small and green with tiny yellow antlers held in erect clusters. They are pollinated by flies and wasps and produce black berries in early spring. With the use of adhesive roots, it winds up trees and trails over buildings and across cliff sides. Its variable dark green or purplish and shiny leaves are extremely shade tolerant, even carpeting the ground in dark woodlands. Thus, it can tolerate many urban situations.

Stinging Nettle *Urtica dioica*
Up to 1.2m (47in)

This plant is famous for its coarse, stinging hairs, which cause minor irritations when brushed against the skin. Flowers appear from June to August, and are composed of tiny, thin, green branched catkins with male and female structures on separate plants. The leaves are toothed, heart-shaped and longer than the stalks, and despite their stinging potential, nettle fibres can be used for making cloth and eaten like spinach. Reliant on phosphorusrich soil, nettles often form extensive patches and can be found growing in woodlands, fens, ditches, the sides of rivers and streams and especially on habitat associated with man, such as farms, gardens, rubbish tips, bonfire sites and abandoned buildings.

Primrose *Primula vulgaris*
20cm (8in)

Gaudy and vulgar forms of this favourite perennial can be found in parks and gardens across the land. However, wild primroses are diminishing due to over collection, and their crinkly lanceolate leaves no longer form such dense carpets in woodland clearings, coppices, on hedge banks, old grasslands and wastelands. The pale yellow flowers appear from March through to May often growing to over 3cm (1in) across and are held on long shaggy stalks. There are two forms of flower, 'pin-eyed' and 'thrum-eyed', which can be differentiated by their construction with regard to the stigma and petals.

Buddleia *Buddleia davidii* Up to 3.5m (138in)
A deciduous scrub, its long bent spikes of heavily scented, mauve flowers appear from July onwards and attract a wealth of butterflies. The species is furnished with toothed, lanceolate leaves, which are downy white below. These may be up to 25cm (10in) long and the dead brown flower spikes may persist into the winter on plants outside gardens where it is frequently found. It is able to colonize railway banks, wasteland and quarries, on what appear to be highly precarious sites such as walls, bridges, and other man-made artefacts.

Field Bindweed
Convolvulus arvensis
Up to 60cm (24in)

This attractive perennial has funnel-shaped, pink and white flowers, which appear from June to September. The leaves are arrow shaped and often form dense patches cloaking fences and bushes and twining anti-clockwise around almost any other vegetation. Road verges, railway sidings and wasteland provide a useful habitat for this serious and persistent weed of arable crops and gardens. It is widespread and abundant through most of the British Isles, excluding Northern Scotland and Ireland.

Common Toadflax *Linaria vulgaris*
Up to 40cm (16in)

The yellow, snap dragon-like flowers appear from July to October and are held in dense spikes. Found on grassy banks and wasteland over much of southern and central England, Wales, and as far north as southern Scotland, the Common Toadflax also frequents hedge banks, roadsides and rough grassland, especially on sandy or chalky soils. Like its garden counterparts, it attracts the attention of a variety of bees.

Greater Plantain *Plantago major* Up to 40cm (16in)
This plant has broad, oval leaves with prominent, non-branched veins below flowers, which have greenish spikes that support a tight cluster of conspicuous, purple antlers on the long, narrow, leafless stalk. The flowers, which appear from June to August, are self-pollinated. Once picked, the stalk can be folded in on itself, which allows the flower heads to be catapulted. It is common on disturbed ground and is nearly always found on paths, tracks and in gateways where it is remarkably tolerant of trampling. Wasteland, roadsides and our lawns provide an alternative habitat.

Goosegrass or Cleavers
Galium aparine
Up to 1.2m (47in)

This straggly, weedy annual will be well known to all those who have strayed from footpaths, because it readily and irritatingly clings to clothing. Its leaves are held in whorls of six to eight and are narrow, broader at the tip, and end in a minute bristle. These fine prickles coating its stems and fruit are successful in effecting the species' dispersal, often aided by the rambler. Goosegrass flowers from June to August and its inconspicuous, white insect-pollinated clusters can be found throughout the British Isles. Hedgerows, ditches, scrubs, streams, riverbanks, scree, shingle beaches and coarse waste ground provide a habitat for this abundant weed.

Common Ragwort *Senecio jacobaea* Up to 1.3m (51in)

Ragwort is one of only five plants notified as a weed by law. It is an often hairless biannual, which has deeply and pinately lobed leaves and large flat-topped flower heads composed of many yellow daisy-like flowers. Its fruits are downy and appear from June onwards, attracting a wealth of insect feeders. It is poisonous to cattle and horses but may be eaten by sheep. Common throughout the British Isles, it appears as a troublesome weed of overgrazed grassland, roadsides, and disturbed ground on all types of soil. On inner city wasteland it often supports large thriving populations of the black and yellow-banded Cinnabar Moth caterpillars.

Groundsel *Senecio vulgaris*
Up to 20cm (8in)

Its flowers, which appear throughout the year, are small, yellow, rayless, cylindrical, and held in short stalks in terminal nodding clusters. The leaves are pinately lobed, hairless above and shaggy below, and its fruits are very hairy or 'dandelionesque' and attract numerous animals. Throughout the British Isles, disturbed ground of any kind whether in gardens or waste places is sure to provide suitable habitat for this untidy and variable annual weed.

Coltsfoot *Tussilago farfara* 15cm (6in)

This species is easily distinguished by its white, hairy, purplish stems, which have overlapping, fleshy scales reaching right up to the flower heads that taper into the stalk. The flower is superficially dandelion-like but has both disc and ray florets. When in fruit, a large amount of white pappus is formed. Coltsfoot leaves are large, finely black-toothed and heart-shaped, initially whitish, they later turn green above. Pollinated by bees and flies, the flowers appear in March and April, and during the night and dull weather, they close up. Coltsfoot favours clay soils, but is also a common plant of roadsides, hedge banks, fields and wasteland.

Daisy *Bellis perennis*
Up to 12cm (4¾in)

Probably the most familiar British wild flower, Daisies are famed for their association with lawns. In natural vegetation, however, they are found on short grassland and can endure a wide range of soils. Tolerant of both heavy grazing and trampling, they flower from January to October when their pretty, white and yellow, composite heads appear on leafless stalks. The disc florets found in the centre of the flower

are bright yellow whilst the white ray florets ring these and are often tipped with crimson. If you do not recognize this plant, make a chain of them and beat yourself with it!

Mugwort *Artemisia vulgaris*
Up to 1.3m (51in)

This species is common on much of the wasteland over the British Isles. Its flowers appear between August and September and they are wind pollinated. It is a slightly aromatic perennial often having purplish stems with many branches and its leaves are pinate with only the lower ones stalked. The flowers are small, egg-shaped and composed of yellow-brown disc florets, which are held in dense, many-branched spikes. Thus, they always appear somewhat damaged and incomplete.

Ox-eye Daisy or Marguerite
Leucanthemum Vulgare
Up to 60cm (24in)

Ox-eye Daisies are a familiar sight from May onwards when they decorate many of our roadside verges. These perennials have glossy, dark green, toothed and lanceolate leaves, with solitary flower heads. The flowers are like giant Common Daisies with a yellow disc and conspicuous white ray florets and are pollinated by a great variety of insects. They are common plants of rough grasslands, roadsides, hedge banks and scrubby grassland.

Dandelion *Taraxacum sp*
Up to 30cm (12in)

Like Brambles, Dandelions are an extremely complicated collection of plants with at least 200 'micro species' recorded in Britain. They are well known to all as a tiresome weed of lawns and waste ground and for their 'sun spotting' of meadows and banks in May. Large, composite flowers sprout from a basal rosette of leaves, which are deeply lobed or toothed and often purplish spotted. The stalk produces a milky juice and the fruits are famous as the Dandelion clocks, which are composed of many fluffy parachutes.

London Plane *Platanus acerifolia*
Up to 30m (32⅓yd)

The leaves of this tree are similar to those of the Sycamore. Ivy-shaped, they are hairless below and often marked with dark fungal blotches. The fruits are tiny, spiky balls, held on long, slender stalks and the flowers appear between May and June. However, the bark, which continually peels off to reveal irregular patches of varying shades of black, brown and cream is the London Plane's most distinctive feature. In our inner cities, where many trees are blackened by car exhaust fumes, the glowing white patches of the plane's bark stands out in vivid contrast.

Silver Birch
Betula pendula
Up to 30m (32⅓yd)

This tree is often a pioneer species on wasteland where it prefers sandy or gravel soils. Its bark is papery, peeling and white, the trunk often rugged at the base and the twigs shining brown, often with tiny warts. In spring, long, yellowish male catkins form, whilst, later, the leaves are alternate, pointed, oval and toothed, turning golden in autumn.

Sycamore *Acer pseudoplatanus*
Up to 35m (38⅓yd)

An introduced species, it has separate sex flowers, which hang in long, yellowish-green clusters. These produce winged fruits in dense bunches of keys, which decorate Sycamores in autumn. The bark is smooth and grey, the twig hairless, and the leaves are often blotched with fungus and covered with millions of aphids. It is a pest in woodlands because its effective germination and rapid growth allow it to dominate and prevent other tree species from establishing seedlings.

Xanthoria parietina 8cm (3⅛in)

In shade, the thallus of this lichen is greenish-grey, but in sunlight it becomes brilliant yellow to bright orange. It has long lobes, which often wrinkle and turn up, and its under surface is almost white with a few light-coloured marks. In the centre of the thallus, the *Apothecia* are abundant and consist of orange discs with paler margins. Farm buildings, bird perching sites, rocks and walls all provide a substrate for this species, but it is also one of the most resistant foliate lichens to air pollution, and thus can be found growing on buildings well within city boundaries.

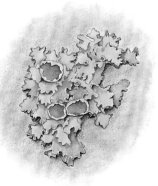

THE WILDLIFE TRUSTS' TOP URBAN SITES

The following is a directory of over 75 of the best urban sites in Britain. The list of sites is divided by local Wildlife Trusts, which are ordered alphabetically. Details on how to get to each site, OS map references, facilities, what to see and when to go are given, but for further information contact the relevant Wildlife Trust.

KEY TO SYMBOLS	
MAMMALS	
BIRDS	
REPTILES & AMPHIBIANS	
INSECTS	
PLANT LIFE	

AVON WILDLIFE TRUST
**The Wildlife Centre, 32 Jacob's Wells Road,
Bristol BS8 1DR
TEL: 0117 917 7270
FAX: 0117 929 7273
EMAIL: mail@avonwildlifetrust.org.uk
WEBSITE: www.avonwildlifetrust.org.uk**

BRANDON HILL NATURE PARK

Address: Jacob's Wells Road, Brandon Hill, Clifton, Bristol
Opening Times: Open at all times.
Facilities: Toilets in park and NCP car park off Jacob's Wells Road. There are nearby cafés in the Triangle and on Park Road.
Location and Access: ST 578728
From M5 Junction 18, take the A4 into Hotwells, then the B4466 to the main entry points off Jacob's Wells Road, Queen Charlotte Street and Berkley Square.
Nearest Train Station: Bristol Temple Meads.
Site Description
Brandon Hill was one of the country's first urban nature parks. It covers 2 ha (5 acres) and supports a variety of habitats, bringing wildlife into the heart of Bristol. A woodland walk reveals an array of woodland wildflowers such as Broad-leaved Helleborine, and among the birds, Song Thrush, Bullfinch and Chiffchaff can be heard if not seen. Other features include a pond, which is home to the Great Crested Newt, and neutral grassland with wildflowers including Betony and Yellow Rattle.

STOCKWOOD OPEN SPACE

Address: The Coots, Stockwood Road, Stockwood, Bristol
Opening Times: Open at all times.
Facilities: Tarmac paths provide access for wheelchairs and a cycle way crosses the reserve.
Location and Access: ST 625693
Take the A37. In Whitchurch turn east at the lights and follow the signs towards Stockwood. There is parking at the end of Stockwood Road.
Nearest Train Station: Bristol Temple Meads.
Site Description
Stockwood Open Space is 53 ha (131 acres) of old farmland made up of meadows, thick hedges and woodlands. There is ancient semi-natural woodland filled in spring with wildflowers such as Bath Asparagus and Bluebells. The scrub land is ideal for nesting Willow Warbler and Whitethroat. During the summer, the neutral grassland is filled with wildflowers such as Pyramidal and Fragrant Orchids. Great Crested Newts and toads are attracted to the ponds

WILLSBRIDGE VALLEY

Address: Willsbridge Mill Visitor Centre, Bath Road, Longwell Green, Avon BS30 6EX
Tel: 0117 932 6885 Fax: 0117 932 9440
Opening Times: Reserve: open at all times. Visitor centre: open at restricted times (ring for details). Exhibition: open on special event days (ring for details).
Facilities: Wheelchair access to most of the valley via Willsbridge Hill; exhibition, car park and toilets.
Location and Access: ST 665708
From the A431 Bristol to Bath road, turn into Long Beach Road. There is a car park on the left.

Nearest Train Station: Bristol Temple Meads.
Site Description
This 8 ha (20 acre) valley is home to a fine example of geological exposure and mammals including otter, fox, Badger and the Greater Horseshoe Bat. The valley's woodlands are at their best in spring when they are full of Bluebells and Nightingale song. Frogs, toads and dragonflies inhabit the ponds. The wildlife garden is an excellent example of how to make your garden a better home for wildlife, and there is also a heritage sculpture trail.

BERKSHIRE, BUCKINGHAMSHIRE & OXFORDSHIRE WILDLIFE TRUST (BBOWT)
**The Lodge, 1 Armstrong Road, Littlemore,
Oxford OX4 4XT
TEL: 01865 775476
FAX: 01865 711301
EMAIL: info@bbowt.org.uk
WEBSITE: www.bbowt.org.uk**

IFFLEY MEADOWS

Address: Off Donnington Bridge Road, Iffley, Oxford, Oxfordshire
Opening Times: Open at all times.
Location and Access:
1.5 miles from the centre of Oxford. From Abingdon Road, Oxford turn onto Donnington Bridge Road and park in Meadow Lane. Walk back over the River Thames and down ramp to towpath. The reserve is immediately on the right hand-side from the bridge to the field beyond the Oxford southern bypass. http://www.bbowt.org.uk/reserves/Iffley-Meadows for map.
Nearest Train Station: Oxford.
Site Description
These ancient water meadows are awash with purple, pink and white snake's-head fritillary flowers in April. Record numbers of the plants thrive here due to careful management. In summer knapweed and great burnet attract dragonflies and banded demoiselle. Listen to the Cetti's and sedge warblers and watch swifts and swallows swoop along the ditches. The site is prone to flooding in winter, which benefits the snake's-head fritillaries.

THE NATURE DISCOVERY CENTRE

Address: Muddy Lane, Lower Way, Thatcham, Berkshire RG19 3FU
Opening times: Nature reserve open at all times. See website http://www.bbowt.org.uk/nature-discovery-centre for details of Visitor Centre opening times and access details.
Nearest train station: Thatcham
Site description: The Nature Discovery Centre is surrounded by a mosaic of different habitats with fantastic wildlife to see all year around. Wander on level footpaths around the lake, visit the internationally-important Thatcham Reedbeds and spend time watching waterfowl from the bird hide. Large groups of wintering wildfowl visit the lake, whilst the reedbeds and hedgerows host fieldfare and redwing; summer visitors to the lake include breeding common terns. Café and shop in the Visitor Centre.

WESTON TURVILLE RESERVOIR

Address: Worlds End Lane, Weston Turville, Buckinghamshire, HP22 5PS

Opening times: Nature reserve open all times. See website http://www.bbowt.org.uk/reserves/Weston-Turville-Reservoir for directions and parking.
Nearest train station: Wendover
Site description: Weston Turville Reservoir, built in 1797 to supply water to part of the Grand Union Canal, is now a species-rich wetland habitat. See hundreds of spring flowering early marsh-orchids followed by wetland flowers. Watch iridescent dragonflies and damselflies from the bird hide. The marshy fen, extensive reedbeds and small woodland create a wildfowl wonderland with teal, tufted duck and the occasional bittern. This is the only known breeding site in Buckinghamshire for the water rail.

THE WILDLIFE TRUST FOR BIRMINGHAM & THE BLACK COUNTRY

16 Greenfield Crescent, Edgbaston, Birmingham B15 3AU
TEL: 0121 454 1199
FAX: 0121 454 6556
EMAIL: info@bbcwildlife.org.uk
WEBSITE: www.bbcwildlife.org.uk

BARR BEACON

Address: Barr Beacon, Beacon Road, Walsall, West Midlands, B74 3TS
Opening times: Open at all times by foot, although the car park is usually locked in the evenings
Facilities: Paths, interpretation and car park
Location and access: SP 061970
Access is from the B4154 Beacon Road with car parking available at both the North and the South of the site.
Nearest train station: Walsall
Site Description
Barr Beacon is at one of the highest points in the West Midlands at 227m above sea level, and offers spectacular panoramic views. The site supports many species such as kestrel and song thrush. Archaeological features include the site of possible Neolithic standing stones, sadly destroyed, and mediaeval ridge and furrow. The most prominent though is a war memorial that stands on the top of the site. It has large expanses of acid grassland with frequent gorse thickets and areas of heathland. There are also mixed broadleaf and coniferous plantations in the southern half.

MOSELEY BOG AND JOY'S WOOD

Address: Moseley Bog Nature Reserve, Yardley Wood Road, Birmingham, West Midlands, B13 9JX
Opening Times: Open at all times.
Facilities: Paths, including all-access walkways, interpretation and car park.
Location and Access: SP 090820
Moseley Bog & Joy's Wood is located in the Springfield area of Birmingham approximately 5.1 km (3.2 miles) south of the city centre. From the B4146 Swanshurst Lane turn right onto Yardley Wood Road. There is another car park further up on the right on Windermere Road.
Nearest Train Station: Hall Green.
Site Description
Here ancient woodland and bog lie alongside more recently created woodland and grassland. An abundance of wildlife includes woodland birds including three species of woodpecker, insects, flowers such as bluebell and wood anemone, and trees including ancient oak and field maple. Once Tolkien's playground, the site also includes archaeological features such as the Bronze Age burnt mounds (a scheduled ancient monument), the remains of Victorian gardens, and an historic dam which once provided a mill pool for the nearby Sarehole Mill.

PORTWAY HILL, PART OF THE ROWLEY HILLS

Address: Portway Hill Nature Reserve, Wolverhampton Rd, Oldbury, B69 2BH (nearest)
Opening times: Open at all times
Facilities: Paths, interpretation. Parking available on adjacent roads.
Location and access: SO 976468
From the A4123 Wolverhampton Road near the KFC, going up the hill to the reserve. Alternatively from an open area off St. Brades Close northwest and via a path skirting the western end of Bury Hill Park and around to the south.
Nearest train station: Sandwell and Dudley or Langley Green
Site Description
Portway Hill, part of the Rowley Hills, looks out over Sandwell, Birmingham and parts of Dudley, and is home to an astounding wealth of grassland wildflowers and butterflies. Created from Dolerite, quarries once dominated the hills. Now there is rolling hillside with flowers including Ox-eye Daisy, Bird's-foot Trefoil and Common Knapweed. On steeper slopes Burnet-saxifrage, Hare's-foot Clover and Silver Hair-grass are found. There are latticed heath and six-spot burnet moths, and butterflies including common Blue, Green Hairstreak and Marbled White. Birds include Skylark, Song Thrush, Linnet and Kestrel.

BRECKNOCK WILDLIFE TRUST

Lion House, Bethel Square, Brecon, Powys LD3 7AY
TEL/FAX: 01874 625708
EMAIL: brecknockwt@cix.co.uk
WEBSITE: www.wildlifetrust.org.uk/brecknock

PWLL Y WRACH

Address: South-east edge of Talgarth, Powys
Opening Times: Open at all times.
Facilities: Interpretation boards, geology trail and car park.
Location and Access: SP 165326
Turn off the A479 in the town opposite the Bell public house, and continue along this narrow road for 1km (½ mile). The entrance is at the end of a 250m (273yd) single track. A small car park can accommodate up to 6 cars. A 594m (650yd) stretch of path to the waterfall is suitable for wheelchair users.
Site Description
This 8.5 ha (21 acre) woodland within the Brecon Beacons National Park contains a variety of interesting flora species including Sessile Oak, Ash, Spindle, Dogwood, heather, Herb Paris and in spring, Bluebell carpets the ground. Other wildlife include birds such as Wood Warbler, Dipper, Grey Wagtail and woodpeckers, and butterflies of which Speckled Wood is one. Two waterfalls and interesting geological features are additional attractions.

WERN PLEMYS

Address: South-east edge of Yslvadgynlais in the Upper Swansea Valley
Opening Times: Open at all times.

Location and Access: SN 788092
From the A4067 (main road through Yslvadgynlais), turn into Pantyffynon Road. There is a small parking area at the end of this Road. Follow the footpath from the gate to the reserve's entrance.
Site Description
Wern Plemys is a small 6 ha (15 acre) woodland with areas of damp, unimproved, herb-rich meadow. The reserve links into a network of local footpaths along the course of an old railway. Noted for its Whorled Caraway, the site is also home to woodland birds including Tawny Owl, Wood Warbler, Chiffchaff and Nuthatch.

CORNWALL WILDLIFE TRUST
Five Acres, Allet, Truro, Cornwall, TR4 9DJ
TEL: (01872) 273939
FAX: 01872 225476
EMAIL: info@cornwallwildlifetrust.org.uk
WEBSITE: www.cornwallwildlifetrust.org.uk/

HALBULLOCK MOOR

Address: Gloweth, Newbridge, Truro
Opening Times: Open at all times.
Facilities: Wide boardwalk running through the woodland with toe rail and passing places.
Location and Access: SW 796444. From the A390 opposite the Treliske Hospital, follow the public footpath which takes you south between housing developments. The reserve entrance is at the bottom of the hill on the left.
Nearest Train Station: Truro.
Site Description
Halbullock Moor is a 4 ha (10 acre) nature reserve. Predominantly made up of wet woodland with a stream in a valley bottom, it is home to a variety of wildlife. Great Spotted Woodpecker, Nuthatch, Treecreeper and Jay are amongst the many bird species seen here as is a wealth of common butterflies such as Orange Tip and Green-veined White. The moor has a fantastic range of woodland flowers, ferns, mosses and lichens, Cornish Moneywort being of particular note. Otters have been recorded in the past.

NANSMELLYN MARSH

Address: Budnick Hill, Perranporth TR6 0DB
Opening Times: Open at all times.
Facilities: Boardwalk, information board, reserves signs and bird-watching hide with ramp (ring Cornwall Wildlife Trust to obtain a key for the hide), car park at rugby club adjacent to site.
Location and Access: SW 762543
Follow the B3285 to Perranporth Rugby Club, which is at the bottom of Budnic Hill. Visitors can park at the rugby club. Access to the reserve is via the gate and ramp to the right of the social club building.
Nearest Train Stations: Truro and Newquay.
Site Description
Saved from becoming a waste tip in the mid 1980s, this 4.5 ha (11 acre) marsh is one of Cornwall's larger reedbeds supporting an abundance of birds, including five species of warbler (Cetti's, Willow, Grasshopper, Reed and Sedge). It also has over 120 species of moths and butterflies such as Hoary Footman, Shore Wainscot, Orange Tip and Common Blue, the rare Desmoulin's Whorl Snail, and attractive marsh plants including the Southern Marsh Orchid, Water Mint and Tussock Sedge. Mammals include Harvest Mouse.

DERBYSHIRE WILDLIFE TRUST
East Hill, Bridge Foot, Belper, Derbyshire DE56 1XH
TEL: 01773 881188
FAX: 01773 821826
EMAIL: enquiries@derbyshirewildlifetrust.co.uk
WEBSITE: www.derbyshirewildlifetrust.org.uk

CARR VALE

Address: Bolsover, Derbyshire
Opening Times: Open at all times.
Facilities: Bird viewing platform, disabled access and car park.
Location and Access: SK 459701
From Chesterfield, follow the A632 to Bolsover and the bottom of Bolsover Hill. At the mini roundabout, turn right and drive up to the county council's Stockley Trail car park. Follow the paths across Peter Fidler Reserve to Carr Vale Flash.
Nearest Train Station: Cresswell.
Site Description
This wetland reserve, formed due to flooding of mining subsidence, is ranked among the top five sites in Derbyshire for birdwatching. Reedbeds and reed marsh attract breeding Reed Warbler and wintering Water Rail. Teal and Wigeon are just two of the large numbers of wildfowl that arrive in winter. In spring and autumn, passage waders and wildfowl include Greenshank and Redshank, which, along with Little Ringed Plover and Lapwing, breed on the reserve.

HILTON GRAVEL PITS

Address: Willowpit Lane, Hilton, Derbyshire
Opening Times: Open at all times.
Facilities: Pond dipping platform for schools and boardwalk suitable for wheelchair users.
Location and Access: SK 249315
From the A50 Derby southern bypass, take the Hilton turn off and turn right at the roundabout. Take the next left into Willowpit Lane and left again to park near the site entrance.
Nearest Train Station: Derby.
Site Description
These disused gravel pits now form an important wetland reserve and an SSSI well known for dragonflies, which include the Ruddy Darter. Woodland and grassy areas make it a good location for Black Poplar, orchids such as Spotted Orchid, and fungi. An impressive 120 species of birds have been recorded including Pochard and Ruddy Duck, Great Crested Grebe and a variety of warblers, tits and all three woodpeckers. The smaller ponds attract Great Crested Newt.

DEVON WILDLIFE TRUST
Cricklepit Mill, Commercial Road,
Exeter, Devon, EX2 4AB
TEL: 01392 279244
EMAIL: contactus@devonwildlifetrust.org
WEBSITE: www.devonwildlifetrust.org

THE OLD SLUDGE BEDS

Address: Adjacent to Countess Wear Sewage Treatment Works, Exeter
Opening Times: Open at all times.
Facilities: A path with boardwalks runs through the site.

Location and Access: SX 952888

From the Countess Wear roundabout, take the A379 towards Dawlish. After 400m (438yd) you pass over the River Exe, turn immediately left into the entrance to the South West Water sewage treatment works and park. Walk along the canal, turning left at the far end.

Nearest Train Stations: Exeter St Thomas and Topsham.

Site Description

The reserve, part of the Exe Estuary SSSI, covers an area of about 5 ha (12 acres). It offers a particularly good example of aquatic plant succession. The plant succession moves through colonies of Water Plantain, Burr-reed and Reed Mace at the edges of the pools to Canary Reed and rushes in areas of swamp; and progresses to willow scrub with Elder, Nettles and Hemlock. The mosaic of water habitats is ideal for aquatic invertebrates. The reserve has become a major stop-off point and feeding area for migrating birds as well as providing an ideal site for breeding species, including Cetti's Warbler. Harvest Mouse is also known to breed on the site.

WARLEIGH POINT

Address: Tamerton Foliot, near Plymouth, Devon

Opening Times: Open at all times.

Facilities: Circular path around the reserve.

Location and Access: SX 447610

Take the B3373 to Tamerton Foliot. The turning for the reserve is signposted and close to Tamerton Foliot Church. The lane follows the tidal creek out towards the estuary. Cars can be parked at the end of this lane near the former railway station at their owners' risk. The reserve entrance is on the far side of the railway bridge.

Nearest Train Station: Plymouth St Budeaux.

Site Description

An SSSI covering 30 ha (74 acres), Warleigh Point is one of the finest examples of coastal oak woodland in Devon. Along with estuarine habitats, there is also a small stream, some acid lushes, a pond, areas of old and new coppice, open glades and scrub. Sessile Oak dominates the woodland and Wild Service tree is present along with Hazel, Guelder Rose and Spindle. Species in the ground flora include masses of Bluebell and Ramsons along with the rare St John's Wort. Insects include the Short-winged Cone Head Cricket. Tawny Owl, Great Spotted Woodpecker and Green Woodpecker breed at Warleigh and there is a large rookery. On the estuary, Redshank and Shelduck are common.

DURHAM WILDLIFE TRUST

Rainton Meadows, Chilton Moor,
Houghton-le-Spring, Tyne & Wear DH4 6PU
TEL: 0191 584 3112
FAX: 0191 584 3934
EMAIL: mail@durhamwt.co.uk
WEBSITE: www.durhamwt.co.uk

RAINTON MEADOWS

Address: Mallard Way, Chilton Moor, Houghton-le-Spring DH4 6PU

Opening Times: The reserve gates are open from 9am – 5pm, Mon – Fri and from 11am to 4pm, Sat, Sun and bank holidays. The coffee shop is open 11am – 4pm every day.

Facilities: Classroom and wildlife garden, displays, exhibitions, coffee shop, toilets (with wheelchair access) and 2 large car parks.

Location and Access: NZ 325486

Off the A690 Durham City to Sunderland road, take the turning signposted from Rainton Bridge on to the B1248. Follow this road for about 0.5km (¼ mile) to the roundabout. Mallard Way is the second turn off on the left. The reserve entrance is signposted.

Nearest Train Station: Durham City.

Site Description

During the early 1990s, the site was worked as a surface mine and then restored to a nature reserve. The 74 hectare site has an excellent network of accessible pathways linking viewing areas that look across the lakes and wetlands, with areas of grassland and woodland completing the habitats on site. The reserve supports over 200 species of bird. Fauna is rich, including brown hare and roe deer, and the wetlands attract numerous dragonflies and damselflies. Butterfly numbers are also high, particularly grassland species.

GLAMORGAN WILDLIFE TRUST

Nature Centre, Fountain Road, Tondu, Bridgend,
Mid Glamorgan CF32 0EH
TEL: 01656 724100
FAX: 01656 729880
EMAIL: glamorganwt@cix.co.uk
WEBSITE: www.wildlifetrust.org.uk/glamorgan

KILLAY MARSH

Address: Ridgeway, Dunvant, Killay, Swansea, South Wales

Opening Times: Open at all times.

Facilities: Boardwalk with wheelchair access into reserve.

Location and Access: SS 597928

From the A4118 Gower Road, turn onto Goetre-Fawr Road. Take the first left turn onto Ridgeway and continue to the bottom of the road and the reserve. There is parking on a quiet street nearby.

Nearest Train Station: Swansea.

Site Description

This urban nature reserve covers 8.6 ha (21 acres). It is covered by a variety of habitats including marsh, fen woodland and grassland which support a range of amphibians including Common Toad and Common Frog, and mammals such as Water Shrew and Water Vole. It is also a historical site for the Marsh Fritillary Butterfly.

LAVERNOCK POINT

Address: Fort Road, Penarth, Vale of Glamorgan

Opening Times: Open at all times.

Facilities: A public footpath runs through the reserve.

Location and Access: ST 182680

From the B4267, turn onto Fort Road signposted Lavernock Point.

Nearest Train Station: Penarth.

Site Description

Unimproved calcareous and neutral grassland flora, which includes Adder's-tongue Fern, a variety of orchids and Dyer's Greenweed, attract a number of butterflies and other insects. The limestone scrub is excellent for viewing birds in winter such as Common Scoter, Manx Shearwater and skuas.

PARC SLIP

Address: Fountain Road, Tondu, Bridgend, South Wales CF32 0EH

Opening Times: Reserve: open at all times. Visitor centre: Monday to Friday 9am to 5pm.

Facilities: Wheelchair access, 3 bird hides, visitor centre, picnic tables, car park and toilets.
Location and Access: SS 880840
From M4 Junction 36, follow the B4281 (to Aberkenfig). Take the turning opposite the Fountain Restaurant. There is a large car park.
Nearest Train Station: Tondu.
Site Description
This 101 ha (250 acre) restored opencast mining area holds a variety of habitats including wetlands, woodland and wildflower meadows. It is an ideal site for butterflies, including a breeding colony of Small Blue, damselflies such as the Scarce Blue-tailed Damselfly, and rare fungi. The birdlife is impressive. Lapwing breed here and other species seen at various times throughout the year include Snipe, Ringed Plover, Green Sandpiper, Green Woodpecker and Skylark. Amongst the flora there is Marsh and Bee Orchids.

GLOUCESTERSHIRE WILDLIFE TRUST
Dulverton Building, Robinswood Hill Country Park, Reservoir Road, Gloucester GL4 6SX
TEL: 01452 383333
FAX: 01452 383334
EMAIL: info@gloucestershirewildlifetrust.co.uk
WEBSITE: www.wildlifetrust.org.uk.gloucswt

CHOSEN HILL

Address: Green Lane, Churchdown.
Opening times: all time.
Location & access: Parking by St. Bartholomew's Church at the top of Green Lane.
Nearest train station: Gloucester.
Site Description
Chosen Hill nature reserve is an imposing sight, located above Churchdown village inbetween Gloucester and Cheltenham. The reserve has excellent views of the Cotswold scarp and contains areas of ancient woodland, scrub and grazed grassland. The oak, ash, hazel and field maple woodland supports one of the most spectacular bluebell displays in the county, as well as other spring flowers such as Common Dog-violet, Yellow Archangel and Wood Anemone. There is a diverse birdlife, with chiffchaff, Great Spotted Woodpecker, Little Owl, Linnet, Nuthatch, Whitethroat and Yellowhammer amongst those breeding. Chosen Hill nature reserve is beautiful in the spring and summer months, perfect for escaping the bustle of nearby Gloucester and Cheltenham.

FROME BANKS

Address: Dr Newton's Way, Stroud GL5 2AP
Opening Times: Open at all times.
Location and Access: SO 851048
Take the A419 in to Stroud and park in Waitrose supermarket. Cross Dr Newton's Way at the pedestrian crossing. Follow the footpath parallel to the viaduct, which takes you down to the banks.
Nearest Train Station: Stroud.
Site Description
The River Frome is a natural river within the Stroud Valleys river system and its bankside vegetation, including Marsh Marigold, Water-mint and Yellow Iris, is a known habitat for otter. The banks are also frequented by Dipper, Grey Wagtail and Kingfisher.

ROBINSWOOD HILL COUNTRY PARK AND LOCAL NATURE RESERVE (OWNED BY GLOUCESTER CITY COUNCIL)

Address: Reservoir Road, Gloucester, GL4 6SX
Opening times: during daylight
Facilities: Toilets, Cafe, picnic benches and carved seats
Location & access: bus route 9 or 7 from Gloucester.
Nearest train station: Gloucester
Site Description
Two-hundred and fifty acres of open countryside with viewpoint, pleasant walks and way-marked nature trails that cover the hill. A traditional orchard has recently been revived and replanted (where appropriate). The summit hosts fabulous views of the Severn Bridge to the South, the Malvern Hills to the North and the Black Mountains to the West.

A terrific walking landscape and enjoyable for animal lovers, the Robinswood Hill is also habitat for many creatures, such as badgers, foxes and birds including Red kites amongst others. Managed by the City Council and home to the headquarters of Gloucestershire Wildlife Trust there are many events throughout the year from practical management, health walks and family fun days.

HAMPSHIRE & ISLE OF WIGHT WILDLIFE TRUST
Beechcroft House, Vicarage Lane, Curdridge, Hampshire, SO32 2DP
TEL: 01489 774400
FAX: 01489 774401
EMAIL: feedback@hiwwt.org.uk
WEBSITE: www.hiwwt.org.uk

FARLINGTON MARSHES

Address: Farlington, Portsmouth
Tel: 01489 774400
Opening Times: Open at all times.
Location and Access: SU 685045
Western car park off the A27/A2030 roundabout (height barrier); eastern access reached by parking at the Broadmarsh Coastal Park.
Nearest Train Station: Hilsea.
Site Description
Farlington Marshes, made up of 125 ha (308 acres) of open grassland, is an SSSI and the Trust's oldest reserve. Although not far from Portsmouth, there is a sense of being out in the wild as one walks the sea-wall footpath. This grazing marsh with lake and ponds is notable for migratory birds, including Black-tailed Godwit, Curlew, Shelduck and Dark-bellied Brent Goose, which in winter flock in their thousands to feed on the harbour's large beds of Eel-grass.

LOWER TEST

Address: Totton, Southampton
Tel: 02380 424206
Opening Times: Reserve open at all times; bird hides generally open from 9am to 5pm.
Location and Access: SU 365145
Turn off the M27 Junction to Redbridge (M271). Follow the signposts to the reserve entrance.
Nearest Train Station: Totton and Millbrook.
Site Description
The River Test flows through the Lower Test Marshes before passing to the sea through Southampton Water. The mixture of freshwater

and saltwater gives a great range of habitats that is reflected in the large numbers of birds and flowering plants. It is one of the best sites in the country to see kingfishers and overwintering wildfowl regularly use this site. Over 450 species of wildflower have also been recorded here, including southern marshland green-winged orchids.

ST CATHERINE'S HILL

Address: Garnier Road, Winchester
Tel: 07831 692963
Opening Times: Open at all times.
Location and Access: SU 484276
The Winchester car park at Garnier Road is off the B3335. The Dongas entrance is off the M3 Junction 9. The site is signposted.
Nearest Train Station: Winchester.
Site Description
An SSSI and a scheduled ancient monument, this site has signs of occupation over thousands of years such as Bronze Age pits. The chalk grassland with Ash and Sycamore wood is rich in wildflowers including Clustered Bellflower and Autumn Lady's Tresses, and a host of scarce butterfly species including Brown Argus and Chalkhill Blue.

HEREFORDSHIRE NATURE TRUST

**Lower House Farm, Ledbury Road, Tupsley,
Hereford HR1 1UT
TEL: 01432 356872
FAX: 01432 275489
EMAIL: herefordwt@cix.co.uk
WEBSITE: www.wildlifetrust.org.uk/hereford**

LOWER HOUSE FARM AND LUGG MEADOWS

Address: Lower House Farm, Ledbury Road, Hereford HR1 1UT
Opening Times: Reserve: open at all times. Farm: Monday to Friday 10am to 4pm.
Facilities: Leaflets available.
Location and Access: SO 533407
Herefordshire Nature Trust headquarters is signposted off the A438 Hereford to Ledbury road. Park in the lane by the Farm.
Nearest Train Station: Hereford.
Site Description
This 48 ha (119 acre) reserve consists of floodplain grassland with woodland, orchard and river habitats. The Meadows are managed today as they were in medieval times as Lammas meadows and are notable for flowers such as Pepper Saxifrage and Snake's-head Fritillary. The River Lugg supports Kingfisher, Sand Martin and otter. It is also notable for aquatic flora and fauna such as Purple-loosestrife.

THE WILDLIFE TRUST FOR LANCASHIRE, MANCHESTER & NORTH MERSEYSIDE

**Cuerden Park Wildlife Centre, Shady Lane Bamber
Bridge. Preston, Lancashire PR5 6AU
TEL: 01772 324129
FAX: 01772 628849
EMAIL: info@lancswt.org.uk
WEBSITE: www.lancswt.org.uk**

FOXHILL BANK

Address: Off Mill Hill, Oswaldtwistle, nr Accrington, Lancashire
Opening Times: Open at all times.
Facilities: Interpretation boards, education worksheets, site leaflet, wheelchair access and small car park.
Location and Access: SD 739278
From Union Road in Oswaldtwistle, follow Foxhill Bank Brow, Badge Brow or Mill Hill, which has car parking for 3 cars, to access the site.
Nearest Train Station: Church & Oswaldtwistle.
Site Description
This 9 ha (22 acre) mosaic of grassland, woodland and scrub, river and reservoirs supports a diversity of flora and fauna including plants such as Knapweed, Ragged-robin and Bistort. Coot, Moorhen, Little Grebe, Mallard and many warblers nest in the undisturbed scrub. Kingfisher, Dipper, Siskin and Blackcap can also be seen. The various habitats attract butterflies and day-flying moths.

WIGAN FLASHES

Address: 225 Poolstock Lane, Worsley Mesnes, Wigan WN3 5JE
Tel: 01204 361847 (Greater Manchester office)
Opening Times: Open at all times.
Facilities: Map available. Footpath network and car park.
Location and Access: SD 585030
Off M6 Junction 25, take the A49 Warrington Road, then the B5238 Poolstock Lane. Turn left at the lights onto Carr Lane to park.
Nearest Train Stations: Wigan and Wigan Wallgate.
Site Description
The Wigan Flashes consist of a 250 ha (618 acre) chain of wetlands, which were formed by subsidence from the mining industry and flooded to create flashes. This mosaic of wetland habitats is of national importance and is notable for reedbeds, birds such as Bittern, Reed Bunting and Reed Warbler, and plants including Dune Helleborine, Round-leaved Wintergreen and Yellow Bird's-nest.

LEICESTERSHIRE & RUTLAND WILDLIFE TRUST

**The Old Mill, 9 Soar Lane, Leicester LE5
TEL: 0116 262 9968
EMAIL: info@lrwt.org.uk
WEBSITE: lrwt.org.uk**

LOUGHBOROUGH BIG MEADOW

Address: Meadow Lane, Loughborough, Leicestershire
Opening Times: Open at all times.
Location and Access: ST 538218
On the edge of Loughborough, alongside Meadow Lane.
Nearest Train Station: Loughborough.
Site Description
This 35.3 ha SSSI is one of the few Lammas Meadows left in England and lies next to the river Soar. The Meadows are the only known site in the county today for the nationally scarce narrow-leaved water dropwort.

LUCAS' MARSH

Address: Brock's Hill Country Park, Oadby, Leicester
Opening Times: Open at all times.
Facilities: Country Park visitors' centre.
Location and Access: SP621998
Park in the Visitors' Centre car park. Go through the gate to the

left of the Centre and walk down the track. The reserve entrance is through a gate on the right.

Nearest Train Station: Leicester.

Site Description

This 1.5 ha Local Nature Reserve is a mosaic of reedbed, ponds, rough grassland, scrub, trees and mature hedgerow that attracts a variety of birdlife and butterflies.

NARBOROUGH BOG

Address: Narborough, Leicester

Opening Times: Open at all times.

Location and Access: SP 549979

The reserve lies between the river Soar, the railway line and the M1 motorway. From the city turn left off the B4114 immediately before going under the motorway and take the track to the sports club. Park in the car park. The reserve entrance is next to the allotments.

Nearest Train Station: Narborough.

Site Description

This 9.2 ha SSSI contains the only substantial deposit of peat in Leicestershire. The main habitats are reedbed, wet woodland and damp meadows. There is a varied ground flora and over 130 species of birds have been recorded.

LINCOLNSHIRE WILDLIFE TRUST

**Banovallum House, Manor House Street,
Horncastle, Lincolnshire LN9 5HF
TEL: 01507 526667
FAX: 01507 525732
EMAIL: lincstrust@cix.co.uk
WEBSITE: www.lincstrust.co.uk**

WHISBY NATURE PARK

Address: Moor Lane, Thorpe on the Hill, Lincoln LN6 9BW

Tel: 01522 500676 or Tel: 01522 696926 (Education Centre).

Opening Times: Park: open at all times. Natural World Centre: daily 10am to 5pm. Car park: 5.45am to 8pm (ring to confirm).

Facilities: Natural World Visitor Centre, easy walking, 7 bird hides, education centre, car park and toilets.

Location and Access: SK 914661

West of Lincoln off the A46 Ring Road. From here the entrance to the nature park and car park are signposted.

Nearest Train Station: Lincoln Central.

Site Description

This nature park is made up of 83 ha (205 acres) of flooded gravel pits, grassland scrub with some wooded areas and a 'Little Heath'. Flowers such as sky blue Harebell and yellow Tormentil grow on this patch of heathland. In summer, Sand Martin are seen around the lake and Common Tern nest on the floating rafts. Kingfisher and Great Crested Grebe can also be seen.

LONDON WILDLIFE TRUST

**Harling House, 47-51 Great Suffolk Street,
London SE1 0BS
TEL: 020 7261 0447
FAX: 020 7261 0538
EMAIL: enquiries@wildlondon.org.uk
WEBSITE: www.wildlondon.org.uk**

CAMLEY STREET NATURAL PARK

Address: 12 Camley Street, London NW1 0PW

Tel: 020 7833 2311 Fax: 020 7833 2488

Email: lwtcamleyst@cix.co.uk

Opening Times: Telephone for opening times and details of the many special events.

Facilities: Disabled facilities, classroom, visitor centre and toilets.

Location and Access: TQ 299835

From King's Cross Underground Station, go along Pancras Road. At the road junction under a railway bridge, turn right into Goods Way. Camley Street is on the left. Follow the signs to the entrance.

Nearest Train Station: King's Cross.

Site Description

The site of a former coal depot, this 1 ha (3 acre) ecological park features a meadow, a wetland area and pond. Bird species include Tufted Duck, Moorhen, House Martin and Kestrel. Visiting birds include Reed Warbler, Siskin and Sparrowhawk. In summer, the meadow is alive with Ox-eye Daisy, Knapweed and Wild Carrot, and insects such as the Common Blue Butterfly and grasshoppers.

CENTRE FOR WILDLIFE GARDENING

Address: 28 Marsden Road, London SE15 4EE

Tel: 020 7252 9186 Fax: 020 7636 7259

Email: lwtwildgarden@cix.co.uk

Opening Times: Telephone for opening times and details of the many special events.

Facilities: Wildlife gardening courses, access for wheelchair users, activity room, exhibitions and toilets.

Location and Access: TQ 338755

From the A2214 East Dulwich Road, turn up Adys Road and follow signs to the centre. From the A2216 Grove Vale, turn up Ondine Road and follow signs. Parking is available on the residential roads.

Nearest Train Stations: East Dulwich.

Site Description

This 0.24 ha (½ acre) award-winning project features a visitor centre, wildflower plant nursery and demonstration gardens, which provide examples on how to support wildlife in a small space. The site is notable for wildflowers, trees, herbs and pond plants. Frogs, newts and garden birds such as Sparrows can also be seen.

THE CHASE

Address: The Millennium Centre, The Chase, Off Dagenham Road, Rush Green, Romford RM7 0SS

Tel: 020 8593 8096 Fax: 020 8984 9488

Email: lwtchase@cix.co.uk

Opening Times: Reserve: open at all times. Millennium Visitor Centre: telephone to check opening times.

Facilities: Visitor centre, exhibitions, trail, car park and toilets.

Location and Access: TQ 515860

From Dagenham East, follow Rainham Road, then turn right into Dagenham Road by Eastbrook public house. There is a signpost to the site along this road.

Nearest Train Station: Dagenham East.

Site Description

This 48 ha (119 acre) nature reserve is next to the Eastbrook End Country Park. The habitats include shallow wetlands, reedbeds, horse-grazed pasture, scrub, which contains hawthorn and willow and woodland with the rare Black Poplar. This site is a haven for birds. The open water attracts Green Sandpiper and Redshank, Teal,

Shoveler and Gadwall. Whitethroat is among the birds found in the scrub and there is the possibility of seeing Kingfisher and Lapwing.

MANX WILDLIFE TRUST
7-8 Market Place, Peel, IM5 1AB
TEL: 01624 844432
EMAIL: enquiries@manxwt.org.uk
WEBSITE: manxwt.org.uk

BALLALOUGH REEDBEDS

Address: Castletown Bypass, Castletown, Isle of Man
Opening Times: Open at all times.
Facilities: Interpretation board, car park and toilets available in Castletown.
Location and Access: SC 258682
The reserve is adjacent to the A5, on the outskirts of Castletown, and between the road and the Douglas to Port Erin Steam Railway.
Nearest Train Station: Castletown.
Site Description
A 1.5 ha (3½ acre) reserve comprising reedbed and meadow habitats, where flora such as False Fox Sedge and Spotted Orchids can be found. There is also a good variety of birds including Reed Bunting, Willow Warbler and Sedge Warbler.

BREAGLE GLEN

Address: St George's Crescent, Port Erin, Isle of Man
Opening Times: Open at all times.
Facilities: Change to read: 'Interpretation board. Toilets at adjacent café / car park area'
Location and Access: SC 196688
From Main Street in Port Erin, turn left towards the beach. Take the next left onto St. George's Crescent by the town clock and follow the road around. The car park is off this road and the reserve lies to the left of the road.
Nearest Train Station: Port Erin.
Site Description
This 0.3 ha (1/2 acre) nature reserve is maintained as a small suburban woodland and stream habitat. The original woodland has been supplemented with additional tree planting and a new hedgerow planted. This now provides a mix of nesting habitat and food resource suitable for a range of bird species that use the site, including Willow Warbler and Spotted Flycatcher, as well as refuge areas for various butterflies and moths including speckled wood and Poplar Hawk-moth.

CURRAGH KIONDROGHAD

Address: Derwent Drive, Onchan, Isle of Man
Opening Times: Open at all times.
Facilities: Interpretation board, pond dipping platform, reserves leaflet available from Trust office, toilets nearby in Onchan.
Location and Access: SC 400782
From Main Street through Onchan, turn onto Church Road. Parking is available on the road. Walk through the village green and the reserve is straight ahead.
Site Description
A former unofficial tip, the rubbish was removed, the pond enlarged and deepened, and a circular boardwalk constructed to create this

0.4 ha (1 acre) nature reserve. The open water, willow scrub, grassland and swamp support a wealth of wetland plants, such as Yellow Iris, Marsh Marigold and Water Mint, and a good variety of birds including Chiffchaff, Goldcrest and Grey Wagtail.

MONTGOMERYSHIRE WILDLIFE TRUST
Collott House, 20 Severn St, Welshpool,
Powys SY21 7AD
TEL: 01938 555654
FAX: 01938 556161
EMAIL: montwt@cix.co.uk
WEBSITE: www.wildlifetrust.org.uk/montgomeryshire

LLYN COED Y DINAS

Address: Sarn Y Bryncaled Roundabout, Welshpool, Powys
Opening Times: Open at all times.
Facilities: A birdwatching hide with wheelchair access and car park.
Location and Access: SJ 223052
Southern end of the Welshpool Bypass. From the roundabout, the entrance is 100m along the A490 into Welshpool.
Nearest Train Station: Welshpool.
Site Description
This 8 ha (20 acre) reserve was created in 1993 after gravel was extracted for the Welshpool Bypass. It comprises a large lake with islands, scrub and a bank for Sand Martins to nest in. Other birds that can be seen here include Lapwing, Little Ringed Plover, Great Crested Grebe, ducks and waders.

PWLL PENARTH

Address: Newtown Water Treatment Works, Llanllwchaiarn, Newtown, Powys
Opening Times: Open at all times.
Facilities: 2 birdwatching hides (access for wheelchair users by arrangement, telephone the Trust) and car park.
Location and Access: SO 137926
Take the B4568 from Newtown to Llanllwchaiarn. Turn by the church and follow the lane to the left for 1.5km (1 mile). Park by the sewage works gates and follow the path.
Nearest Train Station: Newtown.
Site Description
In 1996, the disused sewage treatment settling beds were re-modelled into one lake with islands, surrounded by grass and scrub with an area sown for arable crops and weeds. This 11 ha (27 acre) reserve is notable for birds such as Lapwing, Sand Martin, ducks and waders such as Mallard, and otter.

NORTHUMBERLAND WILDLIFE TRUST
The Garden House, St Nicholas Park, Jubilee Road,
Newcastle-upon-Tyne NE3 3XT
TEL: 0191 284 6884
FAX: 0191 284 6794
EMAIL: mail@northwt.org.uk
WEBSITE: www.nwt.org.uk

BIG WATERS

Address: Sandy Lane, Seaton Burn, Newcastle-upon-Tyne, NE13 7BD

Opening Times: Open at all times.
Facilities: The east end of the site is a country park managed by Newcastle City Council with a network of paths. The west end is a nature reserve with access to tow members' hides and a viewing screen. There is parking on site.
Location and Access: NZ 227734
Drive under the A1 flyover and through Brunswick village. Turn north off Sandy Lane immediately after the residential area. Park at the end of this road. The reserve is signposted.
Nearest Train Station: Newcastle upon Tyne.
Site Description
The largest subsidence pond in the area caused by the collapse of deep mineworkings. Most of the reserve is open water, reedbed and a skirting rim of wet woodland and areas of species rich grassland. Hides (key required, contact the Trust), a bird feeding station and screens allow close observation of wildlife. The reserve has one of the largest colonies of tree sparrows; Great Tit, Blue Tit and Chaffinch are also regular users. Many of the more common water birds are visible, including great crested grebe, mute swan, coot, moorhen, heron and tufted duck. Occasionally, unusual visitors such as water rail and bittern can be spotted. To the East of the reserve the pond is managed by Newcastle City Council as public recreation area. Great crested newt along with a range of damselflies and dragonflies occur here, including large red and Azure Damselfly and Common Hawker.

WEETSLADE COUNTRY PARK

Address: Wide Open, Newcastle upon Tyne, NE23 7LZ
Opening times: Open at all times
Facilities: Car Park, path network and picnic tables.
Location and access: Between Wide Open and Dudley, near the junction B1319 /A189
Nearest station: Newcastle upon Tyne
Site Description
A former colliery site, this reserve has been extensively landscaped to create a wildlife haven on the edge of the city with a hill, grassland, scrub, reedbed and woodland areas. The highest point has good views to the North Sea, the Cheviots and Newcastle city. Prominent on the hill top is the drillhead sculpture, a reminder of its mining past. At the foot of the slopes to the west are three developing reed beds, home to many common damselflies and dragonflies. Many birds are present such as grey partridge, meadow pipit and skylark. At the end of August, flocks of goldfinches can be seen pulling the seeds out of the large stands of teasel.

NORFOLK WILDLIFE TRUST
Berwick House, 22 Thorpe Road, Norwich NR1 1RY
TEL: 01603 625540
FAX: 01603 598300
EMAIL: info@norfolkwildlifetrust.org.uk
WEBSITE: www.norfolkwildlifetrust.org.uk

THORPE MARSHES

Address: Whitlingham Lane, Thorpe St Andrew, Norwich, NR7 0QA
Opening times: Dawn till dusk, every day, all year round
Facilities: footpaths, leaflet, interpretation board, guided walks
Location and access: TL 267083

Located in Thorpe St Andrew, just outside Norwich. To visit the reserve, walk along Whitlingham Lane, at the traffic light junction with Yarmouth Road. At the end of this short road there is a footbridge over the railway line which gives access to the reserve.
Nearest train station: Norwich
Site description:
Located on the eastern fringe of Norwich, this nature reserve has a wonderful mixture of habitats: flower-rich marshes criss-crossed with dykes that are home to many dragonfly and damselfly species, including the rare Norfolk hawker, and the even larger emperor dragonfly. Several species of common butterfly can also be encountered on a good day. The reserve contains a large area of open water: a former gravel working known as St Andrew's Broad. This hosts a variety of waterbirds, particularly in winter, including great crested grebe, pochard, cormorant, grey heron, gadwall and tufted duck. Rarer visitors, such as little egret, are regular.

NOTTINGHAMSHIRE WILDLIFE TRUST
The Old Ragged School, Brook Street,
Nottingham NG1 1EA
TEL: 0115 958 8242
FAX: 0115 924 3175
EMAIL: nottswt@cix.co.uk
WEBSITE: www.wildlifetrust.org.uk/nottinghamshire

KING'S MEADOWS (GRASSLAND)

Address: Situated on the northern edge of the Riverside Industrial Park in the Lenton area of Nottingham.
Opening Times: Open at all times
Location and Access: SK558384.
The reserve is situated on the Lenton Industrial Estate and can be accessed from either Longwall Avenue on the estate or via Birdcage Walk alongside the River Leen (off Lenton Lane). If using satnav, enter NG2 1AE. The reserve entrance is clearly marked by a notice next to the main gate.
Nearest Train Station: Nottingham
Site Description:
King's Meadow is a unique urban nature reserve, created on the former Wilford Power Station site. This small reserve, covering just over 1 hectare, contains an unusually high diversity of habitats within a compact area, supporting a number of rarer plant species such as the Southern Marsh Orchid, Common Spotted Orchid and hybrids. The reserve also has records of invertebrate species found nowhere else in the county.

SELLAR'S WOOD

Address: Sellar's Wood Drive West, Bulwell, Nottingham
Opening Times: Open at all times.
Facilities: Nature trail.
Location and Access: SK 523454
From the A610, take the A6002 Low Wood Road and turn left onto Sellers Wood Drive. Continue over the roundabout onto Sellers Wood Drive West. The reserve is on the right on the corner of Sellers Wood Drive West and Wood Link. Some improvements have been made to facilitate access for wheelchair users, although paths are uneven and wet in places.
Nearest Train Station: Bulwell.
Site Description
This ancient woodland covers 14 ha (35 acres) and is designated as a

Local Nature Reserve and an SSSI. It is a fine example of semi-natural woodland and its varying ecology makes it suitable for a wide range of wildflowers such as Giant Bellflower, Early Purple Orchid, Yellow Archangel and Wood Anemone. The ponds are a haven for amphibians such as frogs, toads, Smooth Newt and various dragonfly species.

WILFORD CLAYPIT (WETLAND) 🐝

Address: Entrance is via Compton Acres, West Bridgford (NG2 7NZ) and Landmere Lane.
Opening Times: Open at all times.
Location and Access: SK571355, SK569354.
2 entrances, one off Landmere Lane, one off Compton Acres in West Bridgford, close to the Apple Tree public house at Compton Acres.
Nearest Train Station: Nottingham
Site Description
This 4.3 hectare SSSI site is a former claypit, with a variety of habitats including marshland, pools, calcareous grassland and areas of scrub and woodland. Waterlogged clays ensure that the marsh is maintained by lime-rich springs, which feed the unpolluted pools and streams. The wide range of plant species includes Bee Orchids and Southern Marsh-orchids, whilst many species of invertebrates can be found on the site including the impressive Blue and Green Emperor Dragonfly.

SHEFFIELD WILDLIFE TRUST

37 Stafford Road, Sheffield, S2 2SF
TEL: 0114 263 4335
FAX: 0114 263 4345
EMAIL: sheffieldwt@cix.co.uk
WEBSITE: www.wildlifetrust.org.uk/sheffield

BLACKA MOOR 🐦🐝

Address: Hathersage Road, Sheffield
Opening Times: Open at all times.
Location and Access: SK 287807
Immediately south of the A625 Hathersage Road. Follow the signs to the reserve.
Nearest Train Stations: Dore and Grindleford.
Site Description
This 181 ha (447 acre) Nature Reserve, which was designated in 2001, lies on the edge of the Peak District National Park. It contains varied areas of woodland, wet woodland and spectacular moorland, and is an important site for woodland birds, especially for upland breeding birds. Species include Curlew, Stonechat, Linnet, Lapwing, Kestrel, Peregrine, Sparrowhawk, Blackcap, Willow Warbler and Tree Pipit. It is also a good site for fungus.

BLACKBURN MEADOWS 🐦🦋🐝

Address: Reserve: Holmes Lock, Steel Street, off Psalters Lane, Rotherham.
Education centre: Blackburn Meadows Education Centre at Magna, Sheffield Road, Rotherham, South Yorkshire S60 1DX
Tel: 01709 723127 Fax: 01709 820092
Email: blackburnmeadows@magnatrust.co.uk
Opening Times: Reserve: open at all times. Education centre: by prior arrangement.

Facilities: Education centre, pond dipping, 2 birdwatching hides and a bird viewing screen, picnic areas, natural artworks, education events and activities, car park and toilets.
Location and Access: SK 415922
Reserve: From the A6109 Meadowbank Road, turn down Psalters Lane, then Steel Street to the car park and reserve entrance. Education Centre: Magna is signposted from the A630 Sheffield Road and is located off Bessemer Way. Ask in reception for the Blackburn Meadows Education Centre.
Nearest Train Stations: Sheffield and Rotherham.
Site Description
Sited on part of an old sewage works, Blackburn Meadows is 21 ha (52 acres), 7 ha (17 acres) of which is open to the public. It is part wetland, part urban savannah, and is important for migrating birds, especially waders. No fewer than 177 species of bird have been recorded, including Greenshank, Snipe, Little Ringed Plover and Gadwall. It is also a good site for butterflies such as Orange Tip, Common Blue and Meadow Brown, the 12 species of dragonfly, including Broad-bodied Chaser, Black-tailed Skimmer, Banded Demoiselle and Emperor, and other invertebrates. Alien and urban flora and fauna such as Michaelmas Daisy, Goat's Rue, Soapwort, Goat Willow, and Rabbits are prolific.

CRABTREE PONDS 🐦🦋🐝

Address: Crabtree Close, off Barnsley Road, Burngreave, Sheffield S5 7AQ
Opening Times: Open at all times.
Facilities: Access for wheelchair users and fishing.
Location and Access: SK 357904
From the city centre, drive along the Wicker, up Spital Hill and along Burngreave Road, which becomes the Barnsley Road (A1635). Turn off into Crabtree Close for the main entrance to the reserve.
Nearest Train Station: Sheffield.
Site Description
Designated in 2001, the 1.5 ha (3½ acre) nature reserve in the heart of inner-city north Sheffield features mixed deciduous woodland and a large urban pond. It is important for bats such as the Pipistrelle, amphibians such as Common Frog, Common Toad, Common Newt and Palmate Newt, and for birdlife including Greenfinch, Great Spotted Woodpecker and Chiffchaff. The site also attracts insects such as dragonflies and damselflies including Brown Hawker and Common Hawker.

SHROPSHIRE WILDLIFE TRUST

193 Abbey Foregate, Shrewsbury, SY3 6AH
TEL: 01743 284280
FAX: 01743 284281
EMAIL: shropshirewt@cix.co.uk
WEBSITE: www.shropshirewildlifetrust.org.uk

GRANVILLE 🦋🐝

Address: Granville Road, Telford, Shropshire, TF2 8PQ
Opening times: Always open
Facilities: Limited disabled access
Location and Access: Follow the signs to Granville Country Park 2 miles north-east of Telford centre. Follow Granville Road for ¼ mile to the car park.
Nearest train station: Telford (limited service to Oakengates)

Site Description

Nature has reclaimed Granville after centuries of industrial activity. It lies at the heart of Telford's green network, with a mosaic of woodland and flower-rich grassland. Old pitmounds have been transformed into flower-rich grassland and heath. An abundance of bird's foot trefoil feeds the caterpillars of Telford's speciality butterflies, the Dingy Skipper and Green Hairstreak.

SHROPSHIRE WILDLIFE TRUST VISITOR CENTRE

Address: The Cut, 193 Abbey Foregate, Shrewsbury SY3 6AH
Tel: 01743 284280
Opening Times: Monday to Saturday 10am to 4.30pm.
Facilities: Wildlife garden, shop, light refreshments, conference rooms and wheelchair access.
Location and Access: SJ 498125
Follow signs to Shrewsbury Abbey off the A5. Enter through a metal gateway next to the lower side of the Abbey Foregate car park.
Nearest Train Station: Shrewsbury.
Site Description
This wildlife garden has evolved from a formal herb garden, reflecting its heritage as part of Shrewsbury Abbey grounds. There is great variety with rambling roses, fruit trees, wild flowers and herb beds. Plants for sale.

SOMERSET WILDLIFE TRUST

34 Wellington Road, Taunton, TA1 5AW.
TEL: 01823 451587
FAX: 01823 451671
EMAIL: somwt@cix.co.uk
WEBSITE: www.wildlifetrust.org.uk/somerset

SUTTON'S POND

Address: Sutton's Ponds are just north of the village of Chilton Trinity, which is on the edge of Bridgwater.
Opening times: Open access, all year round.
Facilities: The bird hide near the car park has disabled access, and some of the paths are suitable for wheelchairs. A second hide is also available to visitors. Caution should be taken near potentially dangerous steep banks and deep water.
Location & access: T 296 396. 7.4 acres (3 ha.) Vehicle access is possible along a metalled minor 'no through road' known as Straight Drove, and a track to the left of this known as Middle Drove.
Nearest train station: Bridgwater
Site Description
The reserve is a worked-out clay pit, two thirds of this reserve consists of open water. In summer, the open water is dominated by water lilies. Great crested grebe have successfully nested on the pond and kingfishers frequent the site, together with several species of dragonfly and damselfly.

STAFFORDSHIRE WILDLIFE TRUST

Coutts House, Sandon, Staffordshire ST18 0DN
TEL: 01889 508534 / 509800
FAX: 01889 508422
EMAIL: staffswt@cix.co.uk
WEBSITE: www.wildlifetrust.org.uk/staffs

BATESWOOD

Address: West of Newcastle-under-Lyme
Opening Times: Open at all times.
Facilities: Viewing butts, marked paths and car park.
Location and Access: SJ 796471
From the A525, take the minor road right just before Madeley Heath. Continue for 1½km (1 mile) and take a right-hand bend. On the left is a house standing above the road on a raised garden. Turn left down a rough track alongside the house.
Nearest Train Station: Stoke on Trent.
Site Description
This 25 ha (62 acre) former opencast site has a variety of habitats including scrub, mature woodland, plantation, wet grassland, meadows, and open water, and there is predominantly easy walking. Bird species include Skylark, Lapwing, Snipe, Grey Partridge and Linnet. Flowers include Ox-eye Daisy, Knapweed, Meadow Vetchling, Cowslip and Common Spotted Orchid. 13 species of dragonfly have been recorded including Ruddy Darter, Banded Demoiselle, Emperor, Black-tailed Skimmer and Black Darter.

DOXEY MARSHES

Address: Wootton Drive, Stafford
Opening Times: Open at all times.
Facilities: 3 birdwatching hides, interpretation boards, 3 nature trails, a good network of paths accessible to wheelchair users and car park.
Location and Access: SJ 903250
Located in the centre of Stafford, adjacent to Sainsbury's. There is car parking at the end of Wootton Drive off Creswell Farm Drive off the A5013 Eccleshall Road. Areas can be flooded in winter.
Nearest Train Station: Stafford.
Site Description
This 121 ha (229 acre) SSSI lies in the floodplain of the River Sow. Subsidence from previous brine pumping has created a habitat of grazing marsh, reedbeds, pools, hedgerows and the largest area of Reed Sweet-grass swamp in the Midlands which supports 250 species of flowering plant such as Marsh Marigold, Yellow Iris and Angelica. Kingfisher and Yellow Wagtail nest on this site and other birds include Snipe, Lapwing, Golden Plover, Shoveler and Skylark. Mammals include otter, Water Vole, bats and Brown Hare, whilst Large Skipper and Small Copper are among the insects.

HEM HEATH & NEWSTEAD WOODS

Address: Nr Trentham, Stoke on Trent, Staffordshire
Opening Times: Open at all times.
Facilities: Paths suitable for wheelchair users, pond dipping platform and car park.
Location and Access: SJ 885411
Take the A5035 road between Longton and Trentham. It is opposite the Trentham Lakes development.
Nearest Train Station: Stoke on Trent.
Site Description
A 30 ha (74 acre) mixed plantation woodland of the mid Nineteenth Century with a distinct semi-natural ancient woodland flora and structure including Bluebells, Wood Sorrel, Figwort, Broad-leaved Helleborine, oak, Sycamore, Beech and Cherry. The woodland attracts birds such as Willow Warbler, Nuthatch and Jay, and Coot and Mallard are lured by the pond. Also seen is the Common Lizard. In autumn the reserve is good for fungi.

SUFFOLK WILDLIFE TRUST

**Brooke House, The Green, Ashbocking,
Nr Ipswich, Suffolk IP6 9JY
TEL: 01473 890089
FAX: 01473 890165
EMAIL: suffolkwt@cix.co.uk
WEBSITE: www.wildlifetrust.org.uk/suffolk**

CARLTON MARSHES

Address: Burnt Hill Lane, Carlton Colville, Suffolk NR33 8HU
Opening Times: Reserve: open at all times. Education centre: Monday to Friday 9am to 5pm, plus Saturday to Sunday in summer.
Facilities: Education centre with visitor facilities including car park and toilets, path suitable for wheelchair users.
Location and Access: TM 508921
From the A146 Beccles road, take a right onto Burnt Hill Lane and follow the brown signs to the car park.
Nearest Train Station: Oulton Broad South.
Site Description
This 45 ha (111 acre) reserve is a miniature broadland grazing marsh, with fens and peat pools, criss-crossed by species-rich dykes. The site is rich in aquatic flora such as Water Soldier, which sinks weighted by chalk to escape the winter ice, and Bogbean. The dykes are also favourite hangouts for dragonflies such as the rare Norfolk Hawker. Birds attracted to the site include Marsh Harrier, which sometimes nests here, waders, Short-eared Owl, Barn Owl, Hen Harrier and Kingfisher. There is also a chance of seeing Water Vole.

LANDGUARD

Address: Viewpoint Road, Felixstowe IP11 8TW
Opening Times: Open at all times.
Facilities: Classroom facilities for booked groups and toilets. The Historic Landguard Fort adjacent to the reserve has additional facilities such as toilets and historical exhibits.
Location and Access: TM 285315
On the A14, follow brown signs for Landguard Point. This takes you right at a crossroads onto Langer Road which becomes Carr Road. Viewpoint Road is on the left. Landguard is just past the Suffolk Sands Caravan Park.
Nearest Train Station: Felixstowe.
Site Description
A fascinating sand and shingle peninsula at the mouth of the River Orwell. The spring and summer bird migrations are spectacular. Breeding birds include Little Tern, Oystercatcher, Ringed Plover, Linnet and Skylark. Wheatear are often present and, in the autumn, large numbers of thrush use the site including Redwing and Fieldfare. The reserve is also notable for rare shingle flowers such as Sea Pea, Yellow Horned Poppy and Stinking Goosefoot.

SURREY WILDLIFE TRUST

**School Lane, Pirbright, Woking, Surrey GU24 0JN
TEL: 01483 795440
FAX: 01483 486505
EMAIL: info@surreywt.org.uk
WEBSITE: www.surreywildlifetrust.org**

DOLLYPERS HILL

Address: Old Lodge Lane, nr Kenley, Caterham, Surrey
Opening Times: Open at all times.
Facilities: Footpath across the reserve.
Location and Access: TQ 315584
From the B2030, access is directly from Old Lodge Lane (north) or Caterham Drive (south), where you can park on the road.
Nearest Train Station: Reedham.
Site Description
This 11 ha (28 acre) reserve is an area of ancient woodland, scrub and chalk downland. The woodland is at its best in spring when it is filled with Bluebell, Wood Anemone, Yellow Archangel, Dog's Mercury and other woodland flowers.

HOWELL HILL

Address: Cuddington Way, Ewell, Surrey
Opening Times: Open at all times.
Facilities: Footpath around the reserve.
Location and Access: TQ 238618
From the Howell Hill roundabout, take the A232 turning to Howell Hill. A footpath leads to the reserve.
Nearest Train Station: East Ewell.
Site Description
This 5 ha (13 acre) reserve comprises old chalk spoil heaps with interesting chalkland flora in various stages of succession. Grassland and scrub areas are developing into woodland. Chalk downland flowers include orchids and Broomrape. There is a variety of insects including Small Blue and Green Hairstreak Butterfly.

THORPE HAY MEADOW

Address: Green Lane, Staines, Surrey
Opening Times: Open at all times.
Facilities: Footpath across the reserve.
Location and Access: TQ 030701
Situated south of Egham Hythe between Devil's Lane (off Thorpe Lea Road B3376) and Green Lane (off Chersey Lane A320). Access to the reserve is by a footpath from Green Lane or Devil's Lane.
Nearest Train Station: Staines.
Site Description
This 6 ha (16 acre) reserve is a small five-sided meadow and the last surviving area of unimproved grassland on the Thames gravel in Surrey. The meadow is bordered by old hedgerows. Meadow grasses and lime-loving flowers such as Cowslip, Yellow Rattle, Clustered Bellflower and Meadow Crane's-bill attract a variety of insects and butterflies including Common Blue and Meadow Brown.

ULSTER WILDLIFE

**TEL: 028 4483 0282
EMAIL: info@ulsterwildlife.org
WEBSITE: www.ulsterwildlife.org**

BOG MEADOWS

Address: Milltown Row, Falls Road, Belfast BT12 6EU
Opening Times: Open at all times.
Facilities: Guided walks, regular events programme, volunteering opportunities and public car park.
Nearest Train Station: Great Victoria Street.

Location and Access: J312726

From M1 Junction 1, take the Donegall Road past the Park Centre and turn left at the end of the road. Continue along the Falls Road. The entrance is signposted to the left. Vehicle access at Milltown Row. Pedestrian access at Milltown Row, roundabout at bottom of Donegall Road and Milltown Cemetery.

Site Description

A 16 ha (40 acres) local nature reserve and remnant of the River Blackstaff floodplain. Ponds, marsh and old meadow habitats attract a great variety of birdlife to the reserve including Tufted Duck, Little Grebe, Willow and Sedge Warbler and Reed Bunting. Snipe are present in high numbers over winter. During the summer months, the reserve attracts Sand Martins, Swifts and Swallows.

WARWICKSHIRE WILDLIFE TRUST

Brandon Marsh Nature Centre, Brandon Lane, Coventry CV3 3GW

TEL: 024 7630 2912

EMAIL: enquiries@wkwt.org.uk

WEBSITE: www.warwickshirewildlifetrust.org.uk

BRANDON MARSH NATURE CENTRE AND SSSSI

Address: Brandon Lane, Coventry CV3 3GW

Tel: 024 7630 2912

Opening Times: Monday to Saturday 9.00am to 4.30pm, Sunday 10.00am to 4.00pm. Free entry for Warwickshire Wildlife Trust Members.

Facilities: Visitor centre, Gift shop, Picnic facilities, Toilets, Disabled toilet, Baby changing. Nature trails, bird watching hides, education centre, sensory garden, tea room and gift shop. A wide range of guided walks, talks and educational activities for all ages throughout the year plus fun days and special events.

Location and Access: SP386761

From the A45-46 northern interchange, take the A45 to London/Rugby. After 200m, turn left into Brandon Lane. The entrance is 1Km (0.5 miles) on the right. Access for wheelchair users.

Nearest Train Station: Coventry.

Site Description

Visitor Centre, cafe and bird hides plus nature trail with disabled access. This large (92.3 ha / 228 acres) Nature Reserve is designated a Site of Special Scientific Interest (SSSI) consisting of freshwater pools created by gravel extraction with reedbeds, willow carr, grassland and woodland. Outstanding for birds including Cuckoo, Kingfisher, Woodpeckers and passage migrants such as Lapwing. Otter numbers have increased in recent years. Grass Snakes and 29 species of butterfly have been recorded on the reserve.

LEAM VALLEY AND WELCHES MEADOW

Address: Leam Valley Local Nature Reserve, Newbold Comyn, Leamington Spa

Opening Times: Open at all times.

Facilities: Birdwatching hide.

Location and Access: SP 330658

Welches Meadow is on the south side of the River Leam and the reserve is on the north where it forms part of Newbold Comyn, which is signposted in Leamington. Car parking at the end of Newbold Terrace East and off Radford Road, off the A425.

Nearest Train Station: Leamington.

Site Description

Both sites have been declared Local Nature Reserves. Habitats include flood meadows, woodland, wetland, marsh and dry grassland. There is a diverse range of flowers such as Snakes-head Fritillary, butterflies, trees, shrubs and birds including Meadow Pipits, Skylarks, Barn Owls and Kingfishers. There are also dragonflies and damselflies such as White-legged Damselfly.

PARKRIDGE CENTRE

Address: Brueton Park, Warwick Road, Solihull, Warwickshire, B91 3HW

Tel: 0121 704 0768

Opening Times: April - October 10.00am - 4.30pm, Nature Area closes at 4pm, Tea Room closes at 4.30pm. November - March: 10.00am - 4.00pm, Nature Areas closes at 3.30pm.

Facilities: Visitor centre, Cafe, Gift shop, Picnic facilities, Toilets, Disabled toilet, Baby changing.

Location and Access: SP 161787

From the M42 Junction 5, take the A41 towards Solihull town centre. At the first junction, bear left and left again at the roundabout. The car park is on the right. Park in Brueton Pay and Display Park car park up to 3 hours free.

Nearest Train Station: Solihull.

Site Description

Set in the middle of Brueton Park, on the edge of Solihull Town Centre, the Parkridge Centre and its 5.5 acre nature area offer a tranquil setting for environmental education and information, plus a wide variety of nature conservation events and activities throughout the year. A mixture of habitats and features including woodland, grassland and a large pond, along with a small arboretum support a wide range of birds, insects and wild flowers.

WILDLIFE TRUST OF SOUTH AND WEST WALES

Cilgerran, Pembrokeshire SA43 2TB

TEL: 01239 621212

FAX: 01239 613211

EMAIL: wwc@welshwildlife.org

WEBSITE: www.waleswildlife.co.uk

TEIFI MARSHES

Address: Cilgerran, Pembrokeshire SA43 2TB

Tel: 01239 621600

Opening Times: Open all year around, but please check website for opening times.

Facilities: 8 birdwatching hides, 4 with wheelchair access, visitor centre, shop, restaurant/café, classroom, toilets and car park.

Location and Access: SN 188430

From Cardigan, take the A478. After 2km (1 mile) turn east at Pen-y-bryn for Cilgerran. After 1km (½ mile) turn north, following the signpost to the Wildlife Centre. The driveway is the old railway.

Site Description

This 106 ha (262 acre) woodland and wetland reserve holds the second largest reedbed in Wales. There are breeding otters and other mammals include Red Deer, Mink, Red Squirrel and Pipistrelle. Amongst the birdlife are Water Rail, the occasional Bittern, Osprey and a resident population of Cetti's Warbler. Also notable for Brown Hairsteak Butterfly and dragonfly assemblages.

WILTSHIRE WILDLIFE TRUST

Elm Tree Court, Long Street, Devizes,
Wiltshire, SN10 1NJ
TEL: 01380 725670
FAX: 01380 729017
EMAIL: admin@wiltshirewildlife.org
WEBSITE: www.wiltshirewildlife.org

RUSHEY PLATT

Address: Kingshill, Swindon
Opening Times: Open at all times.
Facilities: Boardwalk over marsh, interpretation board, pond with dipping platform.
Location and Access: SU 136835
From the A3102 Great Western Way, access is via the Old Town railway path, the Wilts and Berks canal towpath from Kingshill Road or Redpost Drive reached by passing through the new housing development off Wootton Bassett Road.
Nearest Train Station: Swindon.
Site Description
A 1.04 hectare triangular wetland nature reserve made up of fenland (marsh), ponds and the River Ray. In winter, the reserve is home to wading birds such as Snipe and Little Grebe. In spring it is brimming with frogs, toads and the Smooth Newt. By midsummer, it is a sea of pink Valerian flowers.

SMALLBROOK MEADOWS

Address: Smallbrook Road, Warminster
Opening Times: Open at all times. Small visitor centre (portacabin): open by arrangement.
Facilities: Nature trail leaflet, interpretation boards, car park.
Location and Access: ST 878443
The site is adjacent to Lake Pleasure Grounds just off Warminster town centre. Walk through the Pleasure Grounds from Weymouth Street or from the car park on Smallbrook Road. Access for all visitors including wheelchair users.
Nearest Train Station: Warminster.
Site Description
A series of water meadows and wet woodlands along the rivers Wylye and Were. Habitats include marshy grassland, fen and ponds surrounded by mature hedgerows, scrub and woodland. The 20.49 ha (50.6 acre) reserve contains plants such as Cuckoo Flower, Southern Marsh Orchid and Water Avens, which are now rare due to habitat loss. Butterflies and dragonflies including the Banded Demoiselle, and birds such as the Dipper, Kingfisher, Water Rail, Grey Heron, Snipe, along with Water Voles and a variety of aquatic invertebrates are seen at various times throughout the year.

VINCIENTS WOOD

Address: Off Moss Mead, Frogwell, Chippenham
Opening Times: Open at all times.
Facilities: Interpretation boards, woodland paths; a nature reserve fact sheet is available free from Trust head office and Tourist Centres.
Location and Access: ST898734
From the A350 Hungerdown Lane, turn into Frogwell Road at the roundabout. Take the left turning into Moss Mead, then right. The entrance is at the end of the road, over a small bridge.
Nearest Train Station: Chippenham.

Site Description
5.81 ha (14 acres) of semi-natural ancient woodland, surrounded on three sides by housing, it has a circular woodland path which crosses a medieval woodbank. Trees in the wood include Ash, Maple and Oak, and woodland flowers in spring include Wood Anemone and Bluebell, followed by early Goldilocks Buttercup. Birds include Great Spotted Woodpecker, Blackcap and Goldcrest and butterflies include Speckled Wood. Autumn is good for fungi.

WORCESTERSHIRE WILDLIFE TRUST

Lower Smite Farm, Smite Hill, Hindlip,
Worcester, WR3 8SZ
TEL: 01905 754919
EMAIL: enquiries@worcestershirewildlifetrust.org
WEBSITE: www.worcswildlifetrust.co.uk

IPSLEY ALDERS

Address: Alders Drive, Winyates Green, Redditch, Worcestershire
Opening Times: Open at all times.
Facilities: Circular trail and boardwalk on some areas.
Location and Access: SP 076677
From the A4189 take the first left at the roundabout onto Alders Drive. The site is on the right and there is parking on nearby roads.
Site Description
This 18 ha (44 acre) fen marsh, which is a threatened habitat within the UK, is in the middle of a modern housing development. The waterlogged conditions are created by spring waters saturating the soil. A variety of plants have been recorded including Common Spotted Orchid, Fen Bedstraw and Marsh Stitchwort. Breeding birds include Reed Bunting and Grasshopper Warbler; Snipe feed there in winter.

YORKSHIRE WILDLIFE TRUST

1 St. George's Place, York YO24 1GN
TEL: 01904 659570
EMAIL: info@ywt.org.uk
WEBSITE: www.ywt.org.uk

ADEL DAM NATURE RESERVE

Address: Adjacent to Golden Acre Park, Leeds, LS16 9JY
Website: www.ywt.org.uk/reserves/adel-dam-nature-reserve
Opening Times: Open at all times
Location and Access: SE 272414
If coming by car, park in the Golden Acre car park (off A660) and take the underpass into the park, accessed from the car park of the Mercure Parkway Hotel (LS16 8AG). From the car park follow the footpaths to the bridle path and carry on to the nature reserve entrance. There are wheelchair friendly paths from the car park to the nature reserve and to Marsh Hide, a RADAR key is required by wheelchair users at the nature reserve entrance.
Site Description
A rare combination of wet and dry woodland surrounds a lake and pond frequently visited by kingfishers. Masses of bluebells and fungi in season are also exceptional. Mature native and exotic trees can be found in the mixed woodland, with as many as 36 species. Marsh Hide overlooks the pond and feeding station where you can look for Chaffinch, Nuthatch and Great Spotted Woodpecker. Moorhen,

Coot and Mandarin can be seen raising their families in summer from the Lake Hide. Badgers and Roe Deer visit the site, and a family of foxes have made the nature reserve their home.

PEARSON PARK WILDLIFE GARDEN

Address: Pearson Park Wildlife Garden, Princes Avenue, Hull HU5 2TD
Website: www.ywt.org.uk/pearson-park-wildlife-garden
Opening Times: Open 7 days a week from 10am - 4pm.
Location and Access: TA 0835 3014
The garden is located just off Princes Avenue opposite Westbourne Avenue. Please park in Pearson Park which is adjacent to the Wildlife Garden.
Site Description
Come and discover this secret city garden. The wildlife pond, small stretches of hedgerow, woodland and a wildflower meadow provide homes for a huge variety of urban wildlife. There is plenty to see from garden birds such as blue tits, long-tailed tits and robins to fantastic insects including butterflies, shield bugs, ladybirds, bees and much more. In late spring and early summer, the divine scents of rosemary, oregano and sage waft around the simply stunning mature herb garden. The vegetable growing area is a community favourite as families head over to pick the glut of raspberries and more during the summer.

POTTERIC CARR NATURE RESERVE

Address: Potteric Carr Nature Reserve, Sedum House, Mallard Way, Doncaster DN4 8DB
Tel: 01302 570077
Email: potteric.carr@ywt.org.uk
Website: www.ywt.org.uk/potteric-carr
Opening Times: Nature reserve: 9am - 5pm Kingfisher Tearooms: 10am - 4pm (October - May) 10:30am - 4:30pm (April - September) Car parking: Locked at 5pm - if you would like to stay later, then arrive before 5pm and ask at the front desk. Open seven days a week (check website for Christmas closure times)
Location and Access: SE 589 007
Coming from Doncaster take the White Rose Way (A6182), at the roundabout follow the directions for the M18. Potteric Carr Nature Reserve is signposted. From the A1 (southbound) come off at Junction 35 for the M18, then take Junction 3 towards Doncaster and follow signs for the A6182 (White Rose Way). At the first set of lights you reach, turn right into Mallard Way. Park at Sedum House.
Nearest Train Station: Doncaster; Park & Ride bus service between the train station and nature reserve.
Site Description
Explore the mosaic of habitats and wonderfully diverse wildlife by picking one of the four trails, most of which are suitable for wheelchairs and pushchairs. Discover pockets of woodland and wildflower meadows, and recharge with friends and family at the welcoming Kingfisher Tearooms. Wildlife watching hides allow you to get up close and personal with birds such as woodpeckers and kingfishers, whilst the ponds provide a home for newts, frogs and Whirligig Beetles. The meadows during summer are full of butterflies including Small Tortoiseshell, in comparison to a snow laden winter where you may be lucky enough to spot our overwintering Bitterns or a Roe Deer darting across the white landscape.

WILLOW GARTH

Address: Stocking Lane, Knottingley
Opening Times: Open at all times.
Location and Access: SE 514241
From Kottingley, take the A645, turning left into Trundles Lane, then right into Stocking Lane.
Nearest Train Station: Knottingley.
Site Description
The reserve, covering 6 ha (15 acres) with its areas of open water, marsh and working willow coppice lies alongside the River Aire in an area designed as floodplain. Around the pond, reeds and bulrushes are dominant. The marsh has reed-grass and sedges, Meadowsweet and Hairy Willowherb. Hop, Corn Marigold, Winter-cress and Cuckoo Flower also grow. Passage waders include Greenshank, Redshank, Common and Green Sandpipers. Sedge Warbler is the reserve's most common breeding bird. The Harvest Mouse is the most numerous small mammal, but Bank Vole and Common and Pygmy Shrew are also present.

WOODHOUSE WASHLANDS

Address: Woodhouse Washlands, Retford Road, Woodhouse Washlands, Sheffield
Opening Times: Open all day.
Location and Access: SK 432857
From M1 Junction 31, take the A57 Sheffield Road. At the second roundabout, take the Woodhouse Mill road to the next roundabout. Turn left into Retford Road. Parking is next to the Princess Royal public house or in the car park on Furnace Lane.
Nearest Train Station: Woodhouse.
Site Description
This 66 ha (163 acre) reserve made up largely of grassland straddles the River Rother. A large marsh area and several ponds, ditches, and the cut-off meander of the old river course provide other habitats. The Washlands are important for breeding birds such as Snipe, Lapwing, Skylark and Reed Bunting. Mammals include Fox, Stoat, the increasingly scarce Water Vole, and Harvest Mouse. There are many amphibians too, including Smooth Newt and the endangered Great Crested Newt. Fool's Watercress and Celery-leaved Buttercup reflect the site's traditional use as wet pasture and the old grassland supports Pignut, Great Burnet and Knapweed.

GLOSSARY

Aerator
The agent that brings about exposure to the action of air circulation.

Anathema
A detested thing, which is often cursed.

Arboriculturalist
Someone involved in the propagation and care of trees.

Biodiversity
The variety of life in terms of species in a given area.

Buboes
Inflamed and swollen lymph nodes, typically in the armpits, neck and groin.

Calcareous
Containing calcium carbonate, or, in simple terms, chalky.

Carapace
Thick hard shield that covers the body of crabs, tortoises etc.

Clinker
Coarse stone or ash chippings used to bed in railway sleepers.

Cogeners
Related by blood and descended from a common female ancestor. They are often 'cousins'.

Coleopterist
A person involved in the study of beetles.

Desiccation
The removal of water from something,

the process of dehydration.

Detritivore
An animal that consumes plant debris and dead material.

Digitigrade
Walking with only the toes touching the ground, for example, dogs and cats.

Encyst
To become enclosed by a thick membrane or shell, or cyst.

Eugenics
Propagating or improving a population with the use of controlled breeding for desirable inherited characteristics.

Eyrie
The nest site of the larger rapacious birds, for example, falcons and eagles.

Fructification
The act of bearing fruit, or making fruitful.

Granivore
An animal that, typically, consumes grass seeds such as wheat and barley.

Herbarium
A collection of preserved plant specimens.

Hydrotropic
Water or moisture seeking response by an animal or plant.

Insectivore
An animal or plant that consumes insects.

Intromittent
Entering or inserting.

Lanceolate
Narrow and pointed spear-shaped leaves.

Mycelia
The fruiting part of fungal bodies – typically creeping threads as opposed to mushrooms or toadstools.

Organochlorine
Group of toxic fertilizers or pesticides now outlawed in the UK.

Opisoma
Corresponds to the abdomen of an insect – the large terminal section.

Pappus
Fine feathery hairs attached to seeds of certain plants such as Dandelion.

Passerine
A member of a subdivision of birds known as perching birds.

Pedipalps
Foremost limbs of a spider used for manipulating food and mating.

Perennial
Plant type, which continues to flower year after year.

Petiole
A short stem that supports a leaf from the branch.

Photosynthetic
Plant process of converting water and carbon dioxide into sugars using sunlight.

Phototropic
Light seeking response by an animal or plant.

Pinnate
Having leaflets growing opposite each other, in pairs, on either side of the stem.

Plantigrade
Walking with only the sole of the foot touching the ground.

Propagules
The vegetative reproductive bodies of plants.

Prosoma
Corresponds to the thorax of an insect – between the head and abdomen.

Protozoan
Minute invertebrate, typically amoebae.

Ruderals
Group of plants that colonize disturbed ground.

Sebaceous gland
A gland that produces an oily secretion to lubricate hair and skin.

Suppurating
Discharge of puss from a sore or wound.

Tapetum
The reflective layer on the back of the eyes of certain nocturnal animals.

Thallus
Undifferentiated plant-body of algae, fungi and lichens.

Vanessids
Butterflies belonging to the *Vanessa* family e.g. Peacocks, Red Admirals and tortoiseshells.

FURTHER READING

Asher J., Warren M. & Fox R.
Millennium Atlas of Butterflies in Britain and Ireland
OUP, 2001

Baines C.
The Wild Side of Town
BBC Consumer Books, 1986

Baker N.
Nick Baker's Bug Book
Bloomsbury, 2015

Beebee T. and Griffiths R.
Amphibians and Reptiles
HarperCollins, 2000

Brooks S.
Field Guide to the Dragonflies and Damselflies of Great Britain and Ireland
British Wildlife Publishing, 2000

Bryant J., Small B. and Greenwood P.
Living with Urban Wildlife
Open Gate Press, 2002

Chinery M.
Collins Wildlife Trusts Guide: Butterflies of Britain and Europe
Collins, 1998

Chinery M.
Collins Field Guide to the Insects of Britain and Northern Europe
Collins, 3rd ed., 1997

Corbet G. and Southern H.
The Hand Book of British Mammals
Blackwell Scientific Publication, 3rd ed. 1990

Fitter, R., Fitter A., and Blamey M.
Collins Pocket Guide: Wild Flowers of Britain and Northern Europe
Collins, 1996

Gibbons B.
Field Guide to the Insects of Britain and Northern Europe
The Crowood Press, 1996

Golley M. and Moss S.
The Complete Garden Bird Book
New Holland, 1996

Hammond N.
The Wildlife Trusts Guides to

Garden Wildlife, Birds, Insects, Butterflies and Moths, Trees, Wild Flowers
New Holland 2002

Hammond N.
The Wildlife Trusts Handbook of Garden Wildlife
Bloomsbury, 2014

Mabey R.
Flora Britannica
Sinclair Stevenson, 1996

Moss S. and Cotteridge D.
Attracting Birds to your Garden
New Holland, 2000

Packham C.
Chris Packham's Back Garden Nature Reserve
Bloomsbury, 2015

Porter J.
The Colour Identification Guide to Caterpillars of the British Isles
Viking, 1997

Reader's Digest
Field Guide to Trees and Shrubs of Britain
Reader's Digest Association, 2001

Roberts M.J.
Collins Field Guide to Spiders of Britain and Northern Europe
Collins, 1995

Rushforth K.
Collins Wildlife Trusts Guide: Trees of Britain and Europe
Collins, 1999

Shirley P. and Kirk S.
Urban Wildlife
Whittet Books, 1996

Skinner B. and Wilson D.
Colour Identification Guide to Moths of the British Isles
Viking, 1998

Svensson L., Christie D., Mullarney K.
Collins Field Guide to the Birds of Britain and Europe
HarperCollins, 2001

Wardhaugh A. A.
Bats of the British Isles
Shire Publications, 2000

Wheater C.P.
Urban Habitats
Routledge, 1999

USEFUL ADDRESSES

The Wildlife Trusts
The Kiln, Waterside
Mather Road, Newark
Nottinghamshire NG24 1WT
Tel: 01636 677711
E-mail: info@wildlifetrusts.org
Website: www.wildlifetrusts.org

Wildlife Watch
(Contact details the same as The Wildlife Trusts)
E-mail: watch@wildlifetrusts.org
Website: www.wildlifewatch.org.uk

The Amateur Entomologists' Society
P.O. Box 8774
London SW7 5ZG
Tel: 07788 163951
E-mail: aes@theaes.org
Website: www.theaes.org

The Bat Conservation Trust
15 Cloisters House
8 Battersea Park Road
London SW8 4BG
Tel: 020 7627 2629
Fax: 020 7627 2628
E-mail: enquiries@bats.org.uk
Website: www.bats.org.uk

The British Butterfly Conservation Society
Manor Yard
East Lulworth, Wareham
Dorset BH20 5QP

Tel: 01929 400209
Fax: 01929 400210
E-mail: info@butterfly-conservation.org
Website: www.butterfly-conservation.org

British Dragonfly Society
Secretary: Dr. W. H. Wain
The Haywain
Hollywater Road, Bordon
Hampshire GU35 0AD
E-mail: thewains@ukonline.co.uk
Website: www.dragonflysoc.org.uk

British Trust for Ornithology
The Nunnery
Thetford
Norfolk IP24 2PU
Tel: 01842 750050
Fax: 01842 750030
E-mail: info@bto.org
Website: www.bto.org

Vine House Farm
Deeping St Nicholas,
Spalding,
Lincolnshire,
PE11 3DG
Tel: 01775 630208
Website: www.vinehousefarm.co.uk

Fox Project
The Old Chapel
Bradford Street
Tonbridge
Kent TN9 1AW
Tel: 01732 367397
Fax: 01732 366533
E-mail: vulpes@foxproject.freeserve.co.uk

Website:
www.innotts.co.uk/~robmel/fox-project

FROGlife
Mansion House
27–28 Market Place
Halesworth
Suffolk IP19 8AY
Tel: 01986 873733
Fax: 01986 874744
E-mail: froglife@froglife.org
Website: www.froglife.fsnet.co.uk

The Hawk and Owl Trust
c/o Zoological Society of London
Regents Park
London NW1 4RY
Tel: 020 7449 6601
Fax: 020 7586 2870

The Herpetological Conservation Trust
655a Christchurch Road
Boscombe, Bournemouth
Dorset BH1 4AP
Tel: 01202 391319
Fax: 01202 392785
E-mail: enquiries@herpconstruct.org.uk
Website: www.hcontrst.f9.co.uk

The London Wetland Centre
Queen Elizabeth's Walk
London SW19 9WT
Tel: 020 8409 4400
Fax: 020 8409 4401
E-mail: info@wetlandcentre.org.uk
Website: www.wwt.org.uk

Mammals Trust UK
15 Cloisters House

8 Battersea Park Road
London SW8 4BG
Tel: 020 7498 5262
Fax: 020 7498 4459
E-mail: mtuk@mtuk.org
Website: www.mammalstrustuk.org

Plantlife
21 Elizabeth Street
London SW1W 9RP
Tel: 020 7808 0100
Fax: 020 7730 8377
E-mail: enquiries@plantlife.org.uk
Website: www.plantlife.org.uk

The Royal Society for the Protection of Birds
The Lodge
Sandy
Bedfordshire SG19 2DL
Tel: 01767 680551
Fax: 01767 692365
E-mail: info@rspb.org.uk
Website: www.rspb.org.uk

The Wildfowl & Wetlands Trust
Slimbridge
Gloucestershire GL2 7BT
Tel: 01453 891900
Fax: 01453 890827
E-mail: enquiries@wwt.org.uk
Website: www.wwt.org.uk

PICTURE CREDITS

Competition Winners
Simon Armstrong: P55 (b); Steven J. Brookes: p28 (b), p38 (b), p56 (b), p65 (t); Jenny Bryant: p16 (b); V. Burnside: p86 (t); Bob Deane: p70 (b); A. Dennis: p71 (t); Stephen Greeves: p42 (b); Mr. J. Inglis: p9 (t), p16 (t); Mrs Cynthia Lee: p12 (br); Celine Philibert: p12 (tl); A. Prescott: p92 (t); Mike Rhodes: p58; Dave Robinson: p51 (t); Dr Ian D.C. Shephard: p19 (tcr), p61 (t), p66 (br), p91; Tim Withall: p47 (t), p64, p88 (b).

Bruce Coleman Collection: (Jane Burton) p29 (t), p49, p74, p80 (b), p85; (Derek Croucher) p13; (William S. Paton) p59 (tl), p86 (t); (Andrew Purcell) p11 (tr), p38 (tl); (Kim Taylor) p32 (t), p36, p88 (t), p94, p95 (t), p96.

Sylvia Cordaiy Photo Library: (J. Howard) p39 (t).

David Cottridge: p14 (lc, lb), p14-15 (c), p15 (tl), p18-19 (c), p19 (tr, rt, rb), p22 (lc, lb), p22-23 (c), p23 (tcl,

tr), p32 (b), p83 (tl, tr), p90 (t).

John Daniels: p18 (lc).

Ecoscene/Papilio: p43 (t), p62 (t); (Frank Blackburn) p73 (t); (Anthony Cooper) p60 (t); (Paul Franklin) p84, p97 (t); (Nick Hawkes) p77, p80 (t), p83 (b); (Peter McGrath) p81 (t); (Sally Morgan) p11 (rb), p19 (rc), p75, p79 (r); (Tony Page) p79 (l); (Robert Pickett) p31, p45 (t); (Robin Redfern) p87 (t); (Laura Sivell) p28 (t).

Frank Lane Photography Library: (Gerard Lacz) p99 (t); (W. Meinderts/ Foto Natura) p101 (t, b); (Tony Wharton) p76 (t), p99 (b).

The London Wetland Centre: (Martin Senior/WWT) p105 (t).

Nigel Hicks: p10 (lb).

Natural Image/Bob Gibbons: p78.

Nature Photographers Ltd.: p15 (tcr), p23 (rt); (S.C. Bisserot) p100 (b); (N.A. Callow) p30 (t); (E.A. Janes) p30 (b); (W.S. Paton) p73 (b).

Oxford Scientific Films: (Harold

Taylor Abipp) p33; (Caroline Aitzetmuller) p23 (rc); (Liz and Tony Bomford) p44, p48 (t); (Geoff du Feu) p11 (tcl); (Richard Kirby) p29 (b); (Michael Leach) p81 (b); (Avril Ramage) p41; (Robin Redfern) p47 (b); (Tony Tilford) p11 (tl); (Ian West) p37.

Chris Packham: p17, p24, p26 (t), p40, p51 (b), p54 (b), p56 (t), p57 (t), p68, p71 (b), p76 (b), p100 (t), p102, p106.

RSPB (rspb-images.com): p15 (tcl), p52, p55 (t), p57 (b), p59 (b), p60 (b), p65 (b), p69.

RSPCA:
(Alan Barnes) p34 (t); (Colin Carver) p67; (David Chapman) p35; (John Downer/Wild Images) p63 (t), p89; (Geoff du Feu) p23 (br), p39 (b); (Andrew Forsyth) p48 (b); (Mark Hamblin) p11 (tcr), p62 (b); (Stuart Harrop) p26 (t), (E.A. Janes) p9 (b), p28 (lc), p34 (b), p50; (Brian Lightfoot/ Wild Images) p90 (b); (Susan and Allan Parker) p19 (tl); (T. Phelps/Wild Images) p98 (t); (Andrew Routh) p82; (Margaret Welby) p11 (tr), p38 (t); (Wild Images Ltd) p11 (cr).

Science Photo Library: (CNRI) p20; (Andrew Syred) p21.

Sheffield Wildlife Trust: p103, p124.

Colin Smale: p8, p15 (rc), p29 (c), p42 (t), p61 (b).

Windrush Photos: (Richard Revels) p10 (lt), p14 (lt), p15 (tr), p19 (tcl), p23 (tcr), p27, 43 (b), p87 (b), p93, p95 (b); (Colin Carver) p18 (lt); (Alan Petty) p18 (lb); (David Tipling) p15 (rt, rb), p22 (lt), p23 (tl), p46, p54 (t), p63 (b), p66 (bl), p72, p104 (b); (Peter Cairns) p92 (b); (George McCarthy) p97 (b); (Chris Schenk) p98 (b); (Les Borg) p105 (b).

Artwork Acknowledgements:
Mammal, reptile, amphibian and insect artworks by Sandra Doyle and Stuart Carter; bird artworks by Dave Daly; plant life artworks by Bridgette James except for the following: David Sutton: p120 (tr, br), p121 (br), p122 (tl, bl, cr, br), p123 (tl, cl); Cy Baker: p123 (bl, cr).

t=top; b=bottom; c=centre; l=left; r=right

INDEX

Page numbers in *italic* refer to illustration captions.

AUTHOR'S ACKNOWLEDGEMENTS

Thanks are due to Jo Hemmings and Camilla MacWhannell at New Holland who have shown remarkable patience during an often difficult project; for this I am grateful. Thanks to Sylvia Sullivan for concise and creative editorial endeavours, to Gülen Shevki for the innovative design, and to Barbara Levy for agency duties. I am also grateful to Dusty Gedge for Redstart ravings, to Martin Senior for waxing lyrical about wetlands, to Kerry Law for stuff on Sheffield, and to the local Wildlife Trusts and all the staff at the UK office for their contribution to the urban sites section. Thanks also go to David Mills at the British Wildlife Centre for his help and hospitality, to David Cottridge for his philanthropic photography, to William Cheung at *Practical Photography* for agreeing to run the photographic competition, which subsequently provided some of the pictures for this book, and, finally, to Joe McCubbin for typing my illegible scrawl.